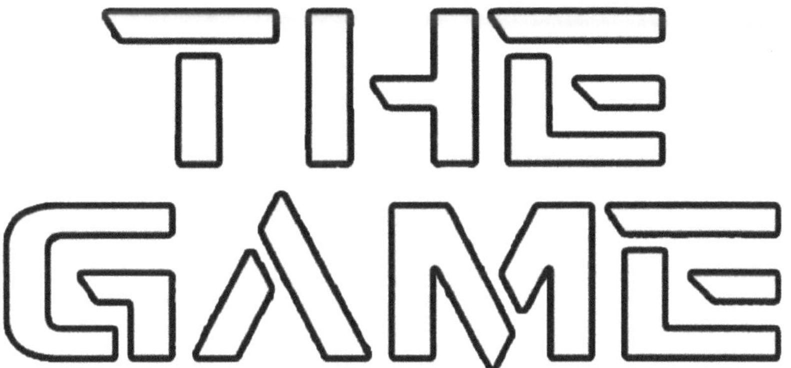

LET'S PLAY SPACESHIPS

DELUXE EDITION

SCOTT RIVERA

The Game: Let's Play Space Ships
Copyright © 2024 by Scott Rivera

ISBN: 979-8894790961(sc)
ISBN: 979-8894790978(hc)
ISBN: 979-8894790985(e)

All rights reserved. No part of this publication may be reproduced, distributed, or transmitted in any form or by any means, including photocopying, recording, or other electronic or mechanical methods, without the prior written permission of the publisher and/or the author, except in the case of brief quotations embodied in critical reviews and other noncommercial uses permitted by copyright law.

The views expressed in this book are solely those of the author and do not necessarily reflect the views of the publisher, and the publisher hereby disclaims any responsibility for them.

The Reading Glass Books
1-888-420-3050
www.readingglassbooks.com
fulfillment@readingglassbooks.com

the Game "let's play spaceships"
Table of contents

THE GAME

The Book that you are about to read is not only a story but has evolved from playing a childhood game of just playing "spaceships". First with my cousin then with my brothers and gradually the game formed and took shape and then it took on a life of its own. Therefore, where our game left off this story was born. As my brother said, "For every imaginative idea that has ever been thought. There is a universe out there somewhere that actually exists, and perhaps created by that thought."

And Now the Beginning...

Chapter 1 *"The leap from Earth"*

After many years of hard work and careful engineering our dream of creating and building our very own spacecraft was real. My two brothers & I had constructed an elaborate underground laboratory about thirty feet underground. Our laboratory was filled with computers, oscilloscopes, and other various complicated electronic equipment. Our ship was fifty feet long and twenty feet wide. We had set a date for our final launch which was about two weeks from now. Up until now we had taken our ship on test flights all over the Earth. After some of our test flights we would watch the news to see if there were any reports of UFOs/flying saucers. On one of our test's flights, we encountered a group of U.S. military jets. They tried to communicate with us muttering something about, "Unidentified craft you are in direct violation of controlled air space and are in danger of being shot down!" Us, knowing that our ship could out fly them at the drop of a hat. We wanted to put our ship to a real test, so we taunted them a little, merely provoking them to fire a couple sidewinder missiles at us. As the missiles sped towards us at over Mach two.

We sped up just a little and reached Mach sixteen in about .7 seconds. We easily outran the missiles, Like a typical UFO. After leaving them in the dust we laughed all the way back to our underground laboratory.

Our launch date had finally arrived, but a rather unfortunate event happened. As we were finishing up loading supplies into our ship and re-fueling.

Our outside motion detectors picked up the movement of many vehicles like cars & vans approaching. My brother Abran was alerted, so he switched on the visual display to see what kind of vehicles they were, but the vehicles were unmarked.

They stopped in front of the house and a total of about fifty agents leaped out of their vehicles. They had rifles, machine guns and handguns.

They ran up to the front door and kicked it in and stormed into the house! From the relative safety of our laboratory Abran shouted, "Erik, Scott wake up !!! I think the FBI or CIA is here, we got to take off RIGHT NOW!!"

Originally, we had planned on taking off later that night under the cover of darkness. I woke up quick to the frantic sound of Abran's voice yelling at us to wake up. Immediately thoughts of our mother still inside the house where the agents had just invaded, was she safe? To our knowledge our mother was completely unaware of our underground laboratory and ship.

The next thing that ran through my mind was [How did they find us? We were always so cautious.] Then I remembered one day when we were testing our cloaking device for our ship, which was not fine-tuned yet. We discovered that it was emitting large amounts of radiation which meant it was working backwards. It took us about a half an hour to reverse it, but after that it worked beautifully. During that test we also discovered two things, one good and one bad. First the good thing, with the cloaking device on emit cycle it acted as a powerful force shield, and we did not expect that. Then the bad thing is, later that night when we watched the news to find out that communications for pretty much all wireless phone service within a 50-mile radius from our lab was completely disrupted during that test. After we heard that we realized that we probably just gave ourselves away. So, that's when we tentatively set a launch date for the following week.

As the agents stormed through the house they found and captured our mother. Then they brought in an armored personnel carrier with a battering ram on it. All three of us were fully aware of the situation.

Erik exclaimed, "What about mom we can't just leave her, those bastards got her!" I said, "Erik there's nothing we can do right now! We have to take off otherwise our mission and our dreams will be ruined! We'll come back for mom someday. Besides, I'm pretty sure that they won't hurt her?"

We had planned for this type of emergency in case the police, FBI, CIA, or any other invasion happened to our lab. We built in a self-destruct detonator into our laboratory just in case this kind of situation should happen, it was the absolute last resort.

It took all three of our passwords entered into the main computer to initiate the self-destruct. Once it was set, we had 7 minutes til' destruction, 7 was Abran's favorite and lucky number. With much trepidation each one of us entered our passwords into the computer followed by a confirmation. The countdown had started 6:59,6:58…etc.

Then I yelled, "There's no stopping it NOW! We better get out of here!" Erik hit the button to open the Big and heavy launch bay doors overhead, they will take a little more than a minute to open all the way. We ran up the ramp into the ship. Erik closed the door after we were inside. Immediately Abran and I sat down at the controls, and with a single touch the panels in front & all around us lit up and came to life. Then I started charging up the main engines. Abran engaged the heart and soul of our ship, which is the Anti-Gravity-Nuclear-Plasma-Monopole System. With using little effort and energy the ship began to rise like a helium filled balloon shedding all of its Earth-bound weight. The Mono Pole drive had to be engaged very, very gradually otherwise we would take off like a bat out of hell! Like falling away from the Earth. The ship was now exiting up through the launch bay doors, now we were clearly visible.

All the Agents turned to look up at our ship that was now hovering about 300 feet above the ground and about 300 yards away from the house. Some of the agents mindlessly pointed their weapons at us. Using our visual display to look down we could see there was about fifty agents swarming all over the place. During the time it took for us to get in the ship and start the launch process. The armored vehicle had pretty much demolished our beloved house looking for us.

Abran shouted, "Look at what they did to our house, it's history!" So, he almost instinctively aimed one of the side cannons at them and was ready to start shooting. I said, "No WAIT! Mom is down there, look I don't care about those guys, but we can't risk mom's life!"

On the ground one of the commanding agents spoke to us with a Megaphone. He yelled, "You in the craft! you must land immediately! And your mother will not get hurt, do you hear me? We know who you are, and you have 30 seconds to land and surrender!" Erik made a comment, "Hey Scott these guys aren't the FBI or CIA, I don't know who they are?" In the distance our motion detectors picked up more vehicles moving towards us. Erik said, "Hey Scott here comes more of those guys, it looks like they have anti-aircraft guns, and they're gonna try to shoot us down!"

Then Abran shouted," Hey guys there's only about 1:45 left until the lab blows up!" I said, "Alright, Abran turn on the targeting computer, see if you can hit all the agents around mom with the stun-tazer to allow mom a chance to escape." The anti-aircraft vehicles were now in firing range and had targeted our ship. The commanding agent with the Megaphone shouted, "ALRIGHT TIMES, UP YOUR MOTHER IS DEAD!!!" So, the agent standing next to him pointed his pistol at our mother's chest. Mom Shouted, "NO, PLEASE DON'T KILL ME!!" And without any hesitation he fired 5 shots, Killing her! Our Mother fell to the ground. We watched in Absolute HORROR! Abran shouted, "NO, they shot her, she's dead, those guys! KILLED MOM!" So, immediately Abran switched the guns from stun-tazer to kill and started shooting with the assistance from the targeting computer.

Within about 5-10 seconds he had pretty much shot and killed all but eight of the agents. The remaining agents scattered and ran for their lives trying to avoid being shot. Erik yelled out, "Hey guys two important things, first the lab is gonna blow up in about 25 seconds, then the next thing is they just fired a whole bunch of missiles at us. LET'S GET THE HELL OUTTA HERE!" Then Erik engaged the force shields right before the missiles were gonna hit the ship.

Within the safety of our ship, we couldn't feel anything. I said, "Well, we wanted to test our shields thoroughly, now I think they have been." Then I yelled out to Abran "PUNCH IT!!!" So immediately Abran pushed the button to fully engage the main engines. We blasted off literally so fast that we left a crater about 1000 feet wide, then just after that our lab exploded.

By now we were reaching the fantastic speed of Mach 200 [Over one hundred fifty-three thousand] miles per hour, and still gaining speed. From our ship's back window, we watched our former house, and the grandeurs Earth whizzing away from under our feet at a fantastic speed.

We were mesmerized to watch the Earth shrinking into the distance at an ever-increasing speed. Erik made the stunning comment, "Well we've never taken the ship this far or this fast before. Everything seems all good, and besides there's no turning back now anyways!"

I agreed but I was soon thinking [Man, we didn't experience almost any inertia at all after blasting off at that speed. Yet I feel like a part of my consciousness was left behind on Earth.]

It's like the last feeling I had before we launched was now the only thought and objective that took precedence in my mind. That thought was that I hated government & authority which made me have a desire for anarchy and a thirst for revenge on anything showing signs of weakness. I wanted to kill, destroy, and maim for the joy of killing and inflicting pain. The next thought I had was [Are these unfamiliar thoughts a result of our high-speed leap from Earth? Or maybe a result of our Cloaking device, which we didn't have time to test adequately for various hazards.] Then I realized that seeing our very beloved mother murdered before our very eyes. Along with the fact that we had the power to stop it, must be the reason why I was having these thoughts. I guess I had too much faith in humanity that they wouldn't be so cold hearted, and monstrous to kill her. I said in my mind [That is it! I'd have to be psychotic for it not to affect me.] Maybe these foolish ideas of humanity being so compassionate for one another came from the mass media devices like T.V., radio, internet, newspaper etc. All of these are seemingly created for the sole purpose of mass indoctrination & brain washing. Showing everyone how to act, what you are supposed to believe in, who to hate, what to love, and what you're supposed to care about. All of the emotions can be trained and harnessed to control the masses and trick them into doing exactly what you want them to do. With the mass media devices, reality can be cleverly edited and filtered to completely control all thought processes without anyone ever knowing that they are being controlled. All of this nefarious evil intent is innocently disguised with a backing laugh track. Without these mass media devices what kind of independent thoughts would we have?

Also, whatever stupid things we might remember while being entertained by the mass media. What nuance of a memory got replaced by one of these idiotic ideas or subliminal mind control messages, nobody knows? Now my thoughts were simply barbaric, and self-preservation was the most important thing now.

Chapter 2 *"Prelude to discovery"*

Looking through our ship's back window we watched the Earth fading away ever smaller & smaller. Then we passed by the moon, it took less than a minute. We took some close up pictures of the moon as we passed by. Then we sadly watched the Earth & Moon duo become just the size of marbles and watched them get smaller & smaller until they blended into the blackness of space. The sun too started looking like a distant star. Erik wanted for us to pass by as many of the planets in our solar system as we could. I agreed, so he checked the charts that he had calculated before we launched. It took approximately 2 days to a week between each planet we passed. To our surprise our ship's speed continued to increase until we finally seemed to max out at around 19.2 million miles per hour. At this speed we concluded that we weren't traveling fast enough to do much interstellar travel, much less any intergalactic travel. I discussed this fact with my brothers, so we came to the decision that we would just go to the nearest star Proxima Centauri and do some exploring there.

As we passed by each planet, we would intentionally slow down and take some pictures. When we encountered Jupiter, it became much more difficult to control our speed because Jupiter's gravity was stronger than we anticipated. Since our ship's main engine uses gravity in reverse, we had to reduce our main engines down to less than 1%. Once we slowed down enough, we took some pictures and some measurements. We really weren't trying to search for any signs of life unless it was clearly life and pretty obvious.

I mean we weren't planning on landing on some planet and getting all suited up just to go either digging or sloshing through some primordial mud looking for evidence of some ancient bacteria or shit like that. We'd only stop and check it out if we saw something like civilization or satellites, otherwise we'd just keep going. Erik was our navigator because of his math skills and attention to detail. He did some further calculations taking into consideration our ships' limited ability to recycle air, food, and water from waste. He estimated those essential systems could continue operating at nearly 80% for at least 5 years, after that the percentage would rapidly decrease. So, the next determination was that if we wanted to reach Proxima Centauri and have enough time to do any adequate amount of exploring we would have to find some way to increase our speed. So, Erik came up with a possible solution.

Since both Jupiter and Saturn are many times the size and mass of the Earth. We could use those planets' combined gravity to increase our speed perhaps exponentially. We might be able to reach the mind-blowing speed of 500 million miles per hour. Before we left Earth, we had engineered our ship to withstand a maximum speed of around 552 million miles per hour. With the advent of our shields, we might be able to travel even faster than that. We wanted to orbit as close to Jupiter as possible without entering its atmosphere. Then while orbiting Jupiter we could wait until Saturn's orbit lines up with Jupiter's. Then using both planets gravity to attain the maximum speed increase from the slingshot effect. That's the same maneuver that NASA uses to gain speed and redirect their spacecraft. Since we had to wait approximately nine days until the two planets' orbit would align with our calculated trajectory.

During this time, we should be able to fine tune our ship's systems. We wanted to make sure that if we were planning on taking our ship close to its maximum safe speed, it would be ready. During this time, we could also plot the safest course to make sure we don't collide with any celestial bodies or smash into a planet. So, a clear pathway had to be plotted.

Time passed slowly each day, each one of us passed the time differently. I choose to gaze into space looking through the powerful telescope that points out of our ship's rear window. Each waking period I would point the telescope towards a different sector of space, often admiring the dazzling colors of distant nebulas and galaxies. Then I would point the telescope back at Jupiter. Since we were orbiting it so closely you could see amazing close ups of its great storms and swirling clouds with constantly changing beautiful colors. I often found myself searching back towards our sun which by now was just a distant star, searching to see if I could find the Earth in the blackness of space. One time I caught the glimmer of what appeared to be a tiny blue speck, I guess it had to be the Earth, but the vision only lasted a minute then it was gone. I know I thought this before. As I was looking at that distant blue speck, I realized that this might be the last time that we'll ever see the Earth again. This fact alone kind of brought a tear to my eye. Then I convinced myself that I had to be strong for my brother's sake, besides there's really no looking back now. Perhaps after we complete our exploring and survive! We'd once again go back to the Earth, but not on a peaceful visit we would be going there seeking revenge and rightfully so.

Abran on the other hand, since I couldn't peer into his mind to see what he was thinking. The only witness I could bear was that he was so focused on seeking revenge for the cold-hearted murder of our mother. At times he would isolate himself and draw out sketches of exotic weapons that were not traceable like an ice gun or sound wave gun, and other seemingly strange weapons. Among us three, that seemed to be Abran's special talent in our group. Which putting it simply, he had the ability to think very, very outside the box. He would always come up with strange and creative original ideas covering a wide array of topics, mainly to make our ship as efficient as possible. Abran would usually come up with the spark of an original idea then he would tell us about it, and both Erik and I would decide on the feasibility & practicality of his idea. Then we would collaborate on a way to engineer, fabricate, or construct it. Both Erik and I were generally able to put Abran's seemingly very strange ideas into reality. However once whatever idea that Abran had was made real, you could see how Abran had some real genius.

Other activities that Abran would enjoy doing, he would spend time practicing his martial arts like jujitsu & karate. Which tied into his love of breakdancing and other exotic dancing styles.

So, going back to the first subject, a lot of the systems in our ship originated from one of Abran's visionary ideas. The ideas seemed to be either of a divine nature or maybe even demonic, it was impossible to tell. Either way it was more than just a pure accident or coincidence. I didn't say this out loud, but I was sure glad that he was on our side and not against us.

Erik was more predictable than Abran was and is the most analytical amongst us. Once the problem was identified Erik was clearly the master in our group at solving it. If the problem was Mathematical or had a logical basis it was Erik's Specialty. He had a high level of diligence and was always thorough when solving the problem. I'd put him to work drawing blueprints, designing diagrams & vector graphics along with all the other technical aspects we needed done.

A lot of times Erik and I would collaborate on trying to solve problems. We always seemed to work so well together, what I didn't think of, he always did and vice versa. One of the toughest problems we had early on involving the ship's main engine was the fact that it was originally designed to use a nuclear reaction to power the electro-gravitational dynamics. The main problem we kept having was that we couldn't sustain the reaction for more than a tiny fraction of a second.

After all the different attempts at solving the problem Erik went back to the drawing board to apply an old theory that I had created a long time before. That gravity is not actually pulling, instead it's pushing. The theory goes into a lot more detail about this and how it works. Once Erik entered this new variable into his calculations everything just fell into place without any problems and the fact that it worked proved that the theory was correct. Once we initiated the main engine, we noticed that it was converting fuel into energy at a ratio of %99.89 which is astonishing! It was so ultra-efficient using such a small amount of fuel that it almost achieved perpetual motion like status.

The Ninth day had finally arrived and started with our ships' long range scanners alerting us that Saturn has finally come into range and was approaching its perfect alignment with our planned trajectory. The last-minute preparations were now underway, in some ways we were quite apprehensive about what awaits us. On the other hand, we were also relieved that the main waiting period was now over, and the real exploring was about to begin. After what seemed like an endlessly long 12 hours. We were finally at the 30 second countdown then the 10,9,8,7,6,5,4,3,2,1. Then Abran engaged the main engine and strangely we actually started more gradually than we did when we lifted off from Earth. Unlike the launch from Earth within a couple seconds we had picked up much more speed than we did on Earth. Within a few seconds we began to double our speed every second. So, Erik had to help Abran navigate to obtain an extremely specific orbit with Saturn, so we could gain the maximum attainable speed. By the time we encountered Saturn's gravitational field we were already at a velocity of approximately 100 million miles per hour. Erik turned the shields on, as we swung by Saturn, kinda turning towards it instead of away like you would normally do. The more we turned towards Saturn the harder and faster Saturn pushed us away.

At this point we were now picking up tremendous speeds 200 million, 300 million, 400 million, 500 million. Basically, gaining about 50 million miles of speed every couple minutes. It was more than just amazing it was absolutely breathtaking. Looking at the stars through the ship's rear window, as they turned a dark reddish color and got very dim until they faded to black. It was a scary site to see all the stars were now invisible and blackness was all we could see. We finally reached the illusive speed we were seeking. We maxed out around 549.62 million miles per hour. At this speed we could now continue with our plans for exploration. Abran announced, "Within less than an hour we will be exiting our solar system and entering into deep space."

Chapter 3 " Deep Space Odyssey Begins"

After a few days of traveling in deep space Abran observed a strong blip on the long-range gravity detector, it detected a massive disturbance. Once Abran got the full scope of this newfound anomaly he called it out saying, "Hey guys, there's something huge out in front of us about 5.45 billion miles away. Whatever it is, its gravity profile seems to be immeasurably long and skinny." The curious fact that we noticed is, that whatever this thing was it seemed to stretch across all of space as far as our instruments could scan. Once I saw for myself exactly what the instruments were detecting. I immediately decided we must slow down or even stop and check it out. That's when I told my brothers of my decision. So, we had to act quickly because it will take some time for us to slow down or stop. After all the time and effort, it took for us to achieve the speed that we were traveling at. With much hesitation I told Abran to disengage the main engine and put it in reverse. The next thought I had is that [There has to be some massive object out in space that we could use to regain our speed if this turns out to be nothing]. Erik was busy searching through all of the Earth based astronomical charts available trying to find this anomaly. As he figured the charts revealed nothing. In order to get better measurements and a better look at it, we'd of course have to get closer and maybe even stop. Since the Earth based observations had so little information about this region of space. We were now going to chart & map this area for our own purposes. The mapping process was made pretty easy thanks to our dedicated mapping computer, all we had to do was just start recording and that's it.

The reason why I decided to tell Abran to begin now slowing gradually so we could continue to take readings and measurements. Since we didn't know what this thing was, I didn't want to get too close without having some idea of what the hell it was.

We wanted to make sure that this thing wasn't lethal and to see what kind of gravity it had either positive or negative, perhaps even both? Erik and I continued to take readings of the anomaly. After a few moments I became conscious that this newfound anomaly has completely captured our attention.

Shortly after we set off, I remember wondering [Well, what do we do now?] I guess reality kinda set in, I realized there really is no turning back! So, I hoped with all of my might that there really was something out there.

In science class and on T.V. we've always been told that with the universe being so vast that life, Intelligent life had to exist out there somewhere at least in theory. I certainly hoped that they were right.

After we had been slowing for about 4 hours Erik was continuing to take measurements and readings. Erik called out, "Hey Scott come here; you've got to see this!" When I got there Erik pointed out a very intriguing fact about this object, that it seemed to be pulsing at an irregular frequency. Almost like there was something inside or many things inside moving in both directions at an extremely high rate of speed, at least this was Erik's theory of the current facts. The final piece of information about this object was. That as massive as it seemed to be it really didn't have any mass of its own, or you could say that it had a conditional mass.

Which means that sometimes it was there, and sometimes it wasn't. It had an amplitude because it phased in and out of existence. When we slowed down the recording of each pulse to about 1/1,000,000 of a second. The amplitude revealed a doppler effect like something inside was speeding along in both directions at a mind-blowing speed much faster than light. Abran was observing the same screen we were, and he made a stunning comment saying, "It looks like a highway for spaceships just like the ones for cars."

Just after he said that I took a closer look at the screen and said, "I think he's right! I mean by all the measurements & readings we've been taking, what else could it be?" Erik said, "Here comes the latest reading since we're now close enough, it looks like it has both positive and negative gravity."

When we ascertained that data, that kinda confirmed our suspicions that our guess might be actually be true. The next consensus that we all reached is that we'd just get close enough to this thing, within about a 100 million miles away and follow beside it for a while and at the same time studying it.

Presently we had reached a billion miles away and as we kept getting closer. I was becoming more aware of how each of us were reacting to the anxiousness of the impending discovery of possible intelligent life!

Our speed had now diminished to a mere 50 million miles per hour. I figured that we may as well stay at this speed because we were still more than halfway to Proxima Centauri by now. So, if this anomaly turned out to be nothing more than a dead end as far as exploration goes, we'd still be ok to stay with our previous mission.

At our present speed it would take us about 18 hours to get to our predetermined safe distance from the anomaly.

After an anxious 18 hours we finally reached our planned safe distance. We were now close enough so that we could deploy the rest of our scanning equipment. We were all anxious to use our Ultra high resolution camera to see what the hell this thing looked like, that is if we could see anything at all. Up until now the only thing we could see with the naked eye was just the blackness of space.

So, the very thought of getting a really good visual of this thing was tantalizing. It was an agonizing 20 seconds waiting for the camera to deploy & position itself. Then waiting an additional 20 seconds while the auto focus continuously and unsuccessfully tried to focus.

Feeling frustrated, I quickly switched the camera to manual mode and started trying to focus it myself. I too found it rather difficult to achieve a satisfactory focus.

After I finally managed to focus on the only visual discrepancy that was even faintly & subtly apparent. This is the best description that I can provide of what I observed. In the distance I saw an extremely black, and relatively narrow line that seemed to stretch across all of space like a horizon line dividing space in half. When a star happened to be directly behind the line the part closest to the line would appear red shifted and stretched. While the part further away would appear blue shifted and distorted. Then the part at the very center would be completely blacked out. The Black line anomaly was the blackest of blacks that my eyes could possibly perceive. It was somehow noticeably even blacker than the surrounding black space, which I found to be quite disturbing. As I stared at it, I felt as though it was stealing the light out of my retinas. I found it impossible to break my mesmerizing gaze. The black line anomaly had some sort of hold over me. Abran noticed that something was wrong, so he had to shake me in order for me to break my draining gaze. After our glimpse at this thing, we immediately realized that it would be unwise to get too close to this thing, because it seemed obviously unnatural and very powerful.

Erik said, "Hey, I'll mess around with the shield to see if I can make it offer even more protection." After hearing that Abran suggested something that at first sounded completely preposterous when he said, "Hey let's just make two shields, the first one being the cloaking shield, and the second one being the force shield." I gave Erik a little smirk look and said, "Oh, What the hell. So, far most of Abran's ideas seem to be at least plausible." That's when Erik said, "I haven't got a clue on how to do that?"

That's when I had a flash memory of when we were testing the cloaking device and our remedy for the initial problem which was that the cloaking shield would always fade in and out. We thought that it was not getting enough power, or it was the wrong polarity, but after we adjusted both of those, and it still didn't fix the problem. So, I said, "Let's adjust the frequency of the damn thing!"

So, Erik began twiddling the knobs, and when he adjusted the frequency between 5.7-6.0 gigahertz the cloaking shield split itself into two, I said, "Whoops! we went too far!" So, we went back down a little until it was a single shield again, but it made me wonder.

So, I reminded Erik about that particular time, and we used the same approach. Once again Erik turned the knobs to the same frequency 5.7-6.0 gigahertz. Our single shield began to split into two, I told Erik, "Alright that's perfect stop!" After taking some measurements the first Shield closest to the ship was the cloaking shield. The other shield was the protective shield. By adjusting the frequency, we had exquisite control over either one of the shields. We could reverse the arrangement or intensify one more than the other until we were satisfied with the arrangement.

We continued to maintain a safe distance from the black line anomaly, which was now about a 100 million miles away. Even though we were pretty far away from it, the gravitational influence was still quite profound. We noticed that our main engine was operating at 29.2% higher efficiency. It must be a result of the gravitational influence of the black line anomaly. It allowed us to have very exquisite control of our speed on any X,Y,Z axis points.

With our newfound control and protection, I felt more confident to order us to get a little closer, by as much as 10 million miles. The next part of our plan was to follow alongside of this thing for a while. We were very hopeful that we would find something about this thing that might explain its origins. We continued to follow alongside this thing for weeks covering billions of miles, and without any answers. I can only speak for myself, but during those weeks the thought of us finding intelligent life was becoming more & more likely.

We've always seen the depiction of alien life on T.V. or in the movies, maybe they actually had it right?

As I thought of the possibility of encountering intelligent life. What came to mind was the superiority & inferiority complex. Say for an instance that two civilizations encountered each other and were intellectually and technologically similar. They both would be in competition for resources and find each other to be a threat. Which would eventually lead to war and possible annihilation of one or both.

The next scenario might be that if two civilizations encountered each other and one was more advanced. Probably enslavement, plundering, or annihilation of the less advanced civilization would occur.

Then another scenario is if a very advanced civilization encounters a much more inferior one, they might have a master/pet relationship.

Then another scenario is if an extremely advanced civilization, possibly a billion times more advanced exists in the same plain as an inferior one then the two might exist without ever being aware that the other one is even there.

Then one of the last possibilities is that a species that is infinitely more advanced and has moved beyond the need for a physical body and not having a body frees up the thought processes so much more than we could ever phantom. Not having a body means they could exist in all places at all times. Then the only reason for this entity to exist is to create forever. Therefore, creating is love and this entity can only be described as GOD, *maybe I got ahead of myself!*

After I analyzed our situation, I came to realize that if this anomaly is indeed a creation of intelligent life. Then it must be a civilization more advanced than ours, but still towards the lower end of the spectrum that I just described. Therefore, it could be a possible threat. As we discussed the current situation both Abran & Erik seemed to have similar resolve and viewed this anomaly in the same way.

Chapter 4 *"Life in The Multitudes"*

On day #174 of our journey once again the long-range gravity detector blipped. Abran took notice. The detector indicated a change in flow. Abran called out to us, "Hey guys there's something ahead, it's about 970 million miles away." At this distance there is really no way that we could make a visual inspection of this change in the anomaly. At this point I concluded that we should slow down even further to approximately 20 million mph. At that speed we'd arrive at the point of interest in about 48 hours or less. The predicament was the same as when we first encountered the anomaly. Which was that we were in some sort of complex game of cat & mouse, the anomaly being the cat and us being the mouse. The facts at hand were that we didn't know what this thing was and only as much as our instruments could peer into it. Then the other fact was that we were on our own now. There's no one to back us up, and if something went wrong we were awfully far away from home. So, like a good game of chess each move had to be planned very carefully as if we were walking on eggshells. As far as we knew our weapons were probably quite primitive and useless. We only had a 50 cal. machine gun that had computer assisted targeting.

We also had a laser for taking measurements of planets and asteroids etc. However, the laser could probably be used for a weapon. Unfortunately, weapons were one thing that we didn't have very much of. I guess we really didn't think that we would need them very badly, but right now I was kinda wishing that we did.

After 47 hours had elapsed, we now only had about I hour of travel time left until we could see the point of interest. After the last hour passed we were now close enough to actually see the point of interest with our telescope. Erik would be the first one of us to gaze upon the point of interest.

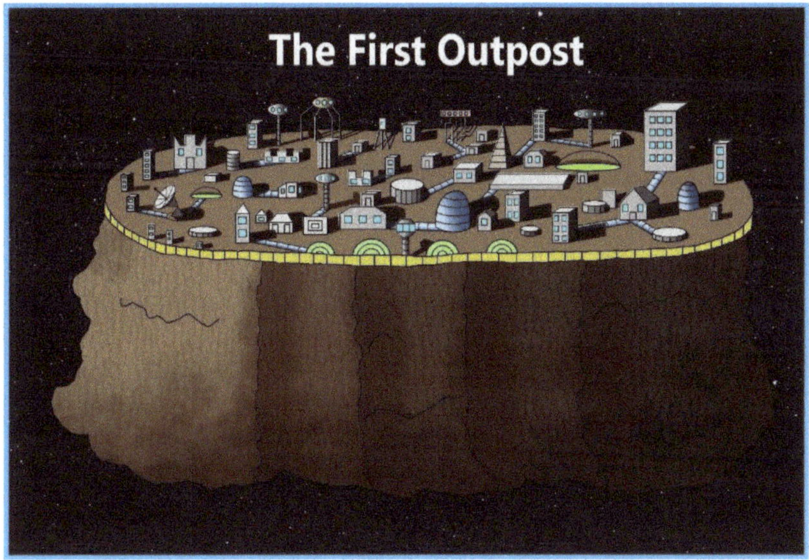

In the dim star light Erik peered at it, and he called out, "Hey guys this is un-believable, you got to see this." As Erik observed, he described, "It looks like spacecraft flying around a small asteroid which looks like it's civilized. It looks like there are buildings and other structures." When it was my turn to look, I saw everything he described. I saw some spacecraft taking off from the asteroid and heading towards what looked like an on-ramp to enter into the black line anomaly. I carefully watched the space craft as they got to a certain point on the ramp then they would just simply vanish. When Abran looked through the telescope he described it as, "Wow! it looks just like a city in space with spaceships getting on and off from an on-ramp to a space highway or something like that."

After hearing Abran say it, both me & Erik agreed that Abran's description sounded right. We all determined that the best and obvious description for the black line anomaly would be to call it the space highway from this point on. While we were so captivated by looking at the asteroid and space highway on-ramp, we weren't paying close attention to our distance from the asteroid. However, we realized this too late. We might have accidentally got too close, immediately we started backing up.

Then unexpectedly our radio light began to flash indicating that we were receiving a transmission which was shocking and horrifying.

We realized that indeed it was too late, and now we have been discovered. Abran asked, "What should we do?" I said ,"I guess answer it, and hope for the best!" Abran reluctantly pressed the audio button, and surprisingly a voice spoke to us in Earth English. Some of the words weren't easy to understand because of the pronunciation but the voice said, "Un-identified craft, please state the nature of your visit" I wasn't exactly sure what to say. So, I just blurted out what I thought sounded good by saying, "We're explorers from the planet Earth, and we're kinda lost." Then the radio voice said, "In that case hold your position, we will send a craft to escort you back to the docking bay." Within minutes a small craft flew out from the asteroid and stopped about ten miles in front of us. Which I felt was a little too close for comfort. Seeing this Abran became very apprehensive saying, "I think their gonna destroy us or something like that. Whatever it is, I don't like it!" In the available light we could see through our telescope what looked like an insignia on the craft that read U.S.P. Inside the cockpit it looked like there were two humanoid pilots.

Then the radio voice once again said, "Un-identified craft please follow us and do exactly what you are ordered to do!" So, we agreed. Then the craft took off back towards the asteroid. The radio voice said, "Follow us." So, we followed them.

As soon as we got back to the asteroid and followed the craft through the entrance into the docking bay our sensors detected a force shield that went up behind us, blocking any escape. This Immediately added to Abran's level of apprehension, and he said, "Great, now we're locked in!"

I tried to comfort Abran by saying, "Try to relax Abran, think about it if they were gonna kill us, they probably would have done it already!" Secretly I was thinking the same thing as he was.

Once we landed in our designated area. The docking bay started pressurizing with air. I thought [Maybe, that's the only reason why the shields had to be up!] but something told me that wasn't it. Once the docking bay was fully pressurized with air, then a door opened towards the front. A small group of soldiers came walking out with what looked like guns in their hands.

Then we heard a loudspeaker announce, "Occupants of this craft you must now exit in a safe and orderly fashion, leave any weapons behind, and don't worry you will not be harmed. Then you will go with these soldiers to a processing center, I reinstate you will not be harmed!"

The moment I heard that I instinctively became very apprehensive about our situation. Abran & Erik also showed that they were getting very worried as well.

A flashback about what happened on Earth went through my mind. Seeing that our situation was kind of a little helpless at this moment, we just had to follow orders. So, Erik reluctantly opened the door then I walked down the ramp first followed by Erik and Abran. We all slowly walked down the ramp. The voice yelled at us, "Stop! and put our hands up! and keep your hands where they can be seen! Now walk slowly towards my voice with no sudden moves!" As soon as we got down to the bottom of the ramp the six soldiers came over to us and told us that they're gonna search us for any weapons. They proceeded to search us and found nothing. After that, the commanding officer told us, "Until we are done with your processing you are under a <u>soft arrest</u> which means you are allowed to move around freely, however you will not be allowed back in your ship for any reason until after your processing is complete. If everything checks out, you will be compensated for your time being detained. Then we will resupply you with whatever we can. Any supplies that we don't have, we will give you a voucher for. Then finally you will be released. For now, come with us." As we left the dock Abran waved goodbye to the ship, we all felt the same way.

The soldiers led us through what seemed like endless hallways and corridors. Abran kept a mental note of all the turns and paths that we were taking in order to find our way back to the ship. This was another one of Abran's talents.

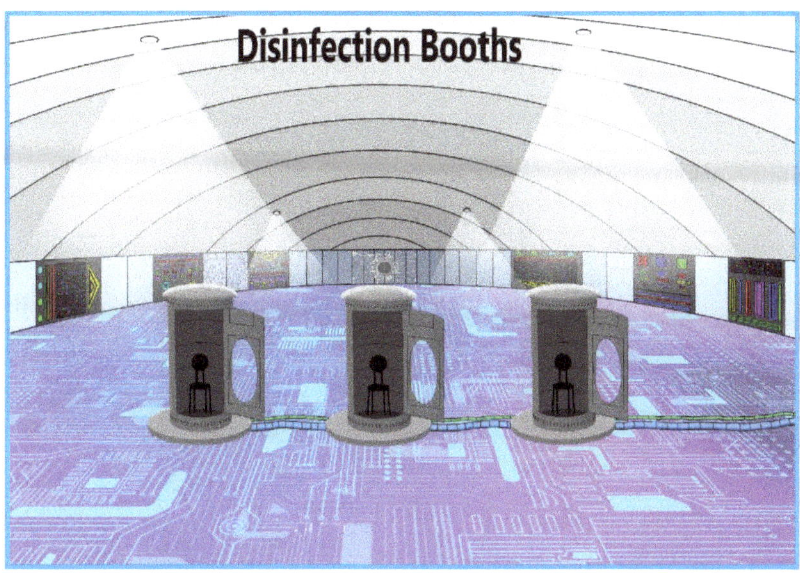

Finally, the soldiers led us to a door which automatically opened to reveal a very large room that looked like some sort of examination room. We were led up to three separate booths in the middle of the room. Inside each booth was a chair, the soldiers told us to get in and sit down then put on the mask. Abran became defensive, the soldier said, "Relax, you guys need to be disinfected just in case you have some kind of infectious disease. We can't risk having you guys get everyone on this outpost sick. So, don't worry this won't hurt a bit."

Once we were seated and our masks were on the operator closed the doors sealing them shut. We could hear through an intercom as the operator said, "Ok, here we go. Oh yeah just breathe normally into your masks and might want to close your eyes."

Then before we knew it the booths started to fill with a thick, yellow-colored smoke that was so thick that you couldn't see anything through it.

Then the operator's voice came on the speaker this time he said, "Don't worry this only takes about a minute." After about a minute went by the operator's voice said, "Ok now stand up and after this next step you're done." So, I proceeded to stand up then as soon as I did the yellow gas began to slightly illuminate and change color to a light purple-violet. My skin began to tingle pleasantly at first. Then the tingle turned to light burn followed by a pretty intense shock which made me jump. I was getting ready to pick up the chair and start trying to break through the glass to get the hell out of there! I know Abran & Erik were thinking the same thing. Then everything stopped and the gas started being sucked out of the booth and was replaced with fresh air. The operator spoke telling us, "Ok, you're all done, sorry about that last step everybody hates that one, but if I told you I'm sure that you would have put up more of a fight. Well anyways the soldiers will take you to the next step of processing, so nice meeting you and goodbye." The doors unlocked and the six soldiers came over and stood in front of the booths. The doors opened and the three of us got out.

The lead officer told us to go through the door over on the side. As we were walking to the door, I looked over to Abran and with a small turn of his eyes I knew that he was thinking the same thing I was, which was <u>Escape!</u> I knew Abran was plotting to escape from the moment we were captured. After exchanging glances with Abran, I turned to Erik to see him slowly nod his head to let me know he was aware that Abran & I were planning some sort of escape. Erik being the more technically minded amongst us. I knew that he'd be looking for any weaknesses in their computer or security systems.

After we went through the door. In the next room we came upon a control panel that looked like I.D. equipment. One officer was seated there, and he said, "Welcome to the next step in your processing which is identification. Here we will be taking your fingerprints, retina scan, and a hair sample for D.N.A. analysis."

After looking at their preliminary screening the officer said, "You're from Earth aren't you?" We replied, "Yes" Then he said, "Somehow I knew that you guys were from Earth. You know this is the furthest away that we've ever picked up any Earthlings before and that makes you guys special."

After we submitted the requested samples for processing then the I.D. officer said, "Well, anyways that's all we need from you guys right now." The lead officer said, "After we get confirmation of your I.D. from the Earth officials using their data bases, then I'm sure that they'll let you go. It shouldn't take longer than 24 hours, maybe sooner. Until then you will be placed into a general holding area, don't worry it'll be pretty comfortable. Then if the Earth officials release you, you will be sent to the Intergalactic bureau of I.D.s and given a new I.D. number, but if the Earth officials request for you guys to be returned, your spacecraft will be confiscated, and you will be sent back to Earth immediately any questions?" We all said, "No."

At this point we all knew two things. One; is that the Earth officials will definitely want us back. Two; is that we had to escape at all costs before we got to the general holding area.

The lead officer said, "Ok come with us." and he pointed towards the direction that he wanted us to go.

I went first, then Erik and then Abran. As we walked, I tried to keep in mind the relative location of our ship. I didn't want them to take us too far away from it that we wouldn't be able to find it. We walked down a long straight hallway that was about 12 feet wide and had intersecting corridors that ran for hundreds of feet in both directions. These hallways didn't seem to have anybody walking down them. I think we all started to realize that now is the best chance to try our escape.

Just then Abran started to act like he was sick, he was having a hard time walking and started to fall down. Abran said, "I think I accidentally inhaled some of that yellow gas in the booth. I feel like I'm gonna pass out!" The soldiers with Abran called for help.

Two of the soldiers that were guarding me & Erik went back to help. The two remaining soldiers stayed back with me & Erik. The soldier who stayed with me pushed me to go forward, so I started walking forward. We came upon an intersection of corridors. I stopped to look back towards Abran. The soldier looked to see what I was looking at.

This was the moment I knew I had to attack. I clenched my left fist together and punched him in the nose as hard as I could! The instant my fist slammed into his nose; blood splattered all over his face and stunned him! Then I jumped-kicked him in the chest, and he fell down in the adjacent corridor, out of sight of the other soldiers! As soon as he was down, I jumped down on him and grabbed him around the neck to begin strangling him! The soldier with Erik saw what was happening and came running to help the fallen soldier!

Erik ran behind him and jumped on top of the soldier before he could pull out his weapon! I strangled the soldier until he went unconscious, then I grabbed his gun and changed the switch on the side from Stun to Kill! Then I shot and killed the soldier that Erik was struggling with, then I shot the soldier that I had strangled!

Me & Erik hid around the corner we could hear Abran keeping the other 4 soldiers occupied as they were trying to help him. The soldiers with Abran seemed completely unaware of what happened to the soldiers that were guarding me & Erik. I peered around the corner and saw Abran. He looked me in the eyes, and I let him have a glance at the gun that I had stolen. As soon as he saw the gun he lunged forward to break free from the soldiers! Then he ran down the hallway away from us as fast as he could! The soldiers shouted, "STOP! or we'll shoot!" Right away I knew this was our only chance. So, both Erik and I ran out to the middle of the hallway and Abran dropped down to the floor to avoid being shot.

That's when me and Erik began shooting as fast as we could, killing all 4 of the remaining soldiers! Once we were sure that all the Soldiers were dead, we stopped shooting. Abran stood up and was unharmed.

We quickly ran over to the dead soldiers and searched them for any items that were deemed to be useful. We grabbed all of their guns and set them to Kill mode. We were relieved that our plan was successful. I said, "That was a little too easy! They don't have very many safety protocols in place to prevent what we just did. Otherwise, they would've used hand or ankle restraints."

We started to run back to the ship, we weren't very far away from it. Just then Erik shouted, "What about the shields, remember they were activated when we docked. We have to find the controls to de-activate them otherwise we can't leave! " Once we reached the dock and saw the ship, we stopped and looked around the corner. We saw a window at the front end of the dock with people behind it.

Erik said, "That kinda looks like a control room, doesn't it?" Abran said, " Yeah, it sure does." I looked around to see how we'd get in there. Suddenly a couple guards entered the dock from a side door, we quickly ducked out of sight.

As we stayed hidden and watched the two guards open the automatic door using what looked like a thumb print reader. Once the guards had entered and the door closed behind them. We walked close to the wall to stay out of sight and crouched to go under the window. Then we pushed a couple of what looked like tool chests next to the door, so we could hide and leap out.

We waited for any other guards to enter the dock so we could use them to gain access to the control room. We didn't have to wait long because we heard some noises coming from just behind the door. We got ready to pounce on whoever exited the door. Then the door opened, and 5 guards exited. We stopped ourselves from leaping out, instead we remained very quiet as they walked further away and exited the dock. The door to the control room stayed open for about 20 seconds and was longer than we thought it would. So, instinctively we bolted through the open door just before it shut!

Once inside we realized that it was basically an elevator which only had two levels. I selected the button for the top level and as the elevator started to move, we got our guns ready and pointed them at the door. Abran said, "Don't shoot if you don't have to because we'll need a technician to de-activate the shields, but after that we can kill them!"

As we waited for the elevator to get to the top, time seemed to stand still for a while. I remember thinking to myself [What have we become three killers? Before this all started, I never hurt anyone, **ever**, nor have my brothers, but now we're talking about killing someone just like taking out the garbage or changing a light bulb, just casual conversation. We've become a bunch of cold-hearted-killers! Is it our fault? Or are we just acting in self-defense?]

Once the elevator stopped time immediately resumed. The door opened and with our guns pointed, we ran out! Just as we speculated it was a control room with dozens of monitors and radar screens. There was only one person in the control room who was a female control operator.

She saw us barge in and looked quite surprised when she said, "Who are you? What are you doing here?" Erik yelled at her, "We want you to deactivate the shields NOW! or we'll kill you!" As we pointed our guns at her then she screamed "OK!! ok please don't shoot!" While all this was happening, I couldn't help but see that she was very beautiful, and I know my brothers were thinking the same thing. I also noticed that her name tag on her shirt said "Kira". As she feverishly entered her password to deactivate the shields.

She said, "The shields are now deactivated, which means they will turn off when your spacecraft leaves the ground. Otherwise, if I turned them off right now you wouldn't be able to board your ship before the vacuum of space would kill you." Abran became annoyed saying, "I don't trust you, to make sure that you're telling the truth we'll take you along as a hostage." Abran grabbed her by the arm and pulled her away from her chair. I was thinking that [Kira, for some reason didn't put up very much resistance.] Now all 4 of us ran back into the elevator. Once it got to the bottom and the door opened, we ran out to the ship. Erik entered the code to open the door. As soon as it was open, we all ran up the ramp. Once inside Abran & Erik immediately started the ships engines. Me and Kira sat in the back seats. Kira said, "You know that once you guys try to take off without authorization, they'll send out fighters to either shoot us down or bring us back. My control room didn't have the ability to give that kind of authorization."

The ship started to lift off and just like Kira had said the dock shields turned off. So, we proceeded slowly towards open space. Abran was an excellent driver; he knew our ship like the back of his hand. Once the ship had completely exited the dock.

Just as Kira warned us, our radio came to life saying, "Craft #1324287 what is the nature of your departure?" I asked Kira, "Where is this person broadcasting from?" She pointed at a small tower with a large window at the top. I saw what she pointed at and showed Abran he turned the ship towards the tower. I sat down in front of the only weapon we had which is a 50 cal. Machine gun.

I remember thinking [I hope this works!] See we've never tested the gun in space before so I had some doubts that it would work. I turned on the computer assisted targeting then when the tower was in sight. I started shooting and I could see glass and body parts flying all about.

During all this Erik had been charging the main engines. So, once the tower was pretty much destroyed, we saw a group of fighters emerging from the docking bay. I cried out to Abran to, "PUNCH-IT" Abran did just that, we zoomed away at over Mach 30 leaving the fighters in the dust.

Chapter 5 " Life as fugitives"

Narrator:

At this point in the story, I realize that I can no longer tell the story from a first-person perspective. I shall take on the omniscient third person perspective. Until it seems reasonable to do so. I am now no longer in this story and shall be able to peer into the minds of every character in this story knowing the thoughts, emotions, past and future all at the same time almost God like, supervising the story.

And now we rejoin the brothers and their hostage where we left off....

The brother's ship whizzed away, putting a growing distance between them and the outpost base. At this point Kira was feeling a little bit more at ease, it seemed that the brothers didn't intend to harm her.

Sometimes you get a gut feeling or bad vibes that you need to listen to when a person is either lying or intends to do you harm. Sometimes these feelings come from what they say or through body language. Kira did not feel any bad vibes from the brothers. In fact, she was beginning to have some ambitions of perhaps even joining them, because deep down she had been yearning for some sort of change in her boring mundane life and this was definitely a change.

After the four of them had been traveling for about an hour Kira started to feel more comfortable, enough to ask Erik, whom she felt a little more willing to talk to than the others. She said, "Excuse me sir, can I use the bathroom?"

Erik said, "Oh yeah, I forgot you don't know our names, I'm Erik and the guy driving is my older brother Abran, and the guy next to you is Scott he's my oldest brother and your name is Kira right just like your name tag says?" She said, "Yea it's Kira, well it's nice meeting all you, so uhh can I still use the bathroom?" Erik realized that the bathroom that they had wasn't really designed for females, they hadn't really planned for this. Scott kinda interrupted saying, "I'll show her where it is and how to use it." Scott led Kira to the bathroom and showed her the toilet, sink, and shower. Then showed her the basics on how to use them. As Scott was leaving her, he said "Well that's the basics on how to use this stuff you know how to do the rest."

As the hours rolled by Abran became more conscience of the fact that they didn't appear to have enough supplies for the 4 of them to last a lot longer. Even taking into consideration the fact that their ship's ability to recycle waste, but fresh supplies are always better.

Abran spoke to Scott about this in quiet tones not wanting to alarm Erik & Kira which the two of them were starting to hit it off really well. Scott was the most practical minded one amongst them. He remembered that near the outpost where they just came from. There was an entry point for the Space highway which he figured was the best way to get around space quickly. Scott thought to ask Kira what she knew about this, so he walked up to Erik and Kira. While Erik was showing Kira some basic ship controls. Scott said, "Sorry to interrupt you two, but we have some important things to discuss. Kira, by now you should know that we don't intend to harm you. Therefore, you must know that we are quite low on supplies and since you probably know this sector of space and we don't, we need your help."

Kira said, "I hate to say it, but other than the outpost where I was stationed the next closest place to get any supplies is all the way in the Reptiless galaxy which is about 8.4 million light years from here. The only way we could ever get there is by taking the space highway. There's an on-ramp for the space highway next to the outpost where I was stationed, but it's probably too dangerous to go back right now. Even then, that's the only way that I can think of how we can get more supplies." Scott said, "So, it's called the space highway… we thought so. To hear you say it just confirms our hypothesis. Well, after listening to you Kira I've decided that we have to go back towards the outpost to get to the entrance of the space highway." That's when Abran spoke saying, " Hey Kira, once we get to the entrance of the space highway how do we get on?" Kira responded, "That's easy all you do is just go in."

Then Erik said, "Hey Kira, you said that it's 8.4 million light years away! How long would that take traveling on the space highway?" Kira said, "A little longer than 8 hours. See the space highway is remarkable because once you enter it. The space highway it carries you along at a million times the speed of light. You can even turn your engines off to save power or you can use them to go even faster. Of course, there's a speed limit according to the laws of physics. One of the limits is that you can't go faster than light speed, you have to go just a little slower. Combining the speed of the space highway along your speed allows you to reach even the most remote locations in the known universe very quickly. Of course, when I say the known universe, I mean places that space highway goes to. There are many uninhabited zones in the universe that the space highway doesn't go to. Anything beyond that is truly unknown."

All three of the brothers were somewhat amazed at Kira's knowledge of the space highway. She also said, "This is pretty much common knowledge. Almost everyone in school is taught this stuff." Scott said, "All that means is that we have to go back to the on-ramp, if we want to go anywhere quickly." Then Kira said, "First of all this ship has force shields, right? Because you're gonna need them when we get on the space highway." Abran said, "Yeah it has force shields" Then Kira said," That's good, so when we get closer to the on-ramp there will be a downloadable map available for Super Quadrant 368 showing the space highway system going through it. Super quadrant 368 is not the biggest or smallest sector of space; it's middle-sized, but it's probably in the top ten of most inhabited." After they all came to the same consensus that they had to take the risk of going back towards the outpost to be able to enter the on-ramp to the space highway, they started to turn the ship around. Abran said, "Well, it'll take a couple hours to get back to the outpost."

Kira was secretly thinking about whether she should try to escape or not, because she barely knew these guys. What if they changed their minds and decided to kill her once they reached the Reptiless galaxy or wherever they decided to go.

Even if they weren't planning on killing her, escape might still be the best decision. If she didn't attempt to escape before they got to the space highway, she might never see her family or friends again? Maybe the brothers would let her at least make contact with them?

After two hours had elapsed Abran announced, "We have the outpost insight, we are only about 1.2 million miles away." Then Erik said, "Ok, I will activate the cloaking shields, the bad thing is that we haven't thoroughly tested them yet, and it kinda uses a lot of energy." Once they got to the million-mile mark Erik engaged the cloaking shields. Then he said, "Let's pray this works." At this point they were still traveling at a little more than a million miles per hour, so that meant about a half hour or less to get to the on-ramp.

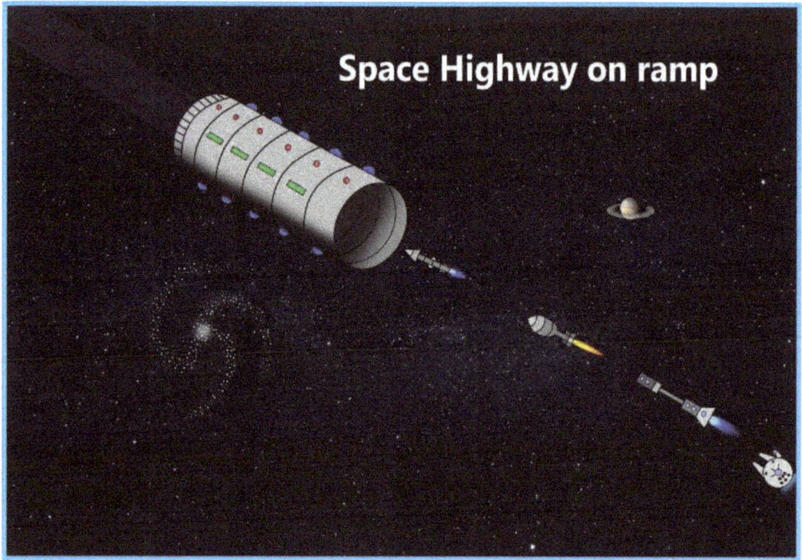

After 25 minutes had passed, and in the dim star light the brothers could now see the on-ramp in the distance. They saw from a distance what looked like tiny craft entering and exiting the ramp. Then they received the map download of super quadrant 368 just like Kira said they would.

They were now close enough and they had slowed down to a mere 10,000 miles per hour. Erik paused with a sly of relief when it was noticed that none of the other crafts seemed to have detected their presence.

So, as they got close to entering the space highway, they watched the other crafts approach the entrance and just simply vanish into a completely black veil. The brothers became a little un-easy about entering. Erik started to give a distance read out as they got closer, starting with 2000 miles then 1000, 500, 400, 300, 200, 100 then finally they entered.

Chapter 6 "The Amazing Space Highway"

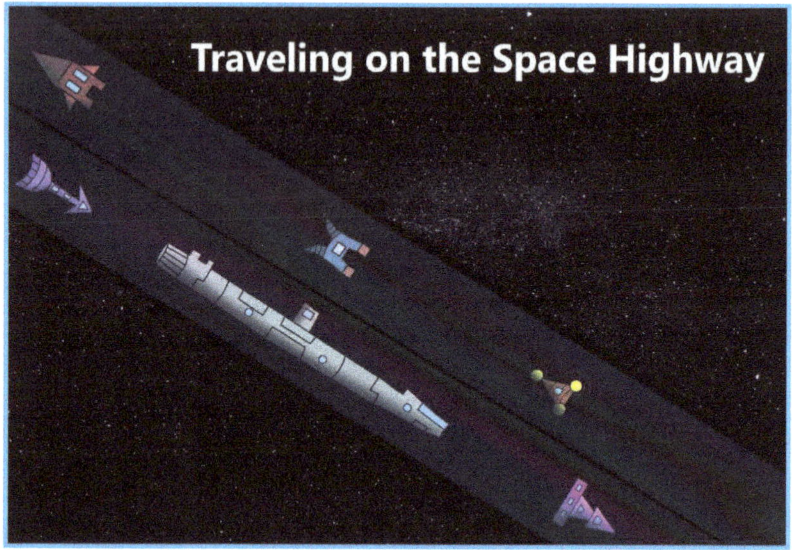

Once they fully entered, the change was un-real. They let out an uncontrollable gasp at the difference in time & speed. It was unlike any other human experience. Kira was the only one of them who was accustomed to this, but the brothers were totally amazed. Each one of the brothers had their own way of coping with this incredible experience. Scott was completely winded, it not only took his breath away, but it ripped the thoughts from his mind. Abran was very fascinated, but like most things he never wanted his feelings known, because that would be a sign of weakness and Abran wouldn't allow that. Erik thought to himself that it felt like falling off a very tall building without a net below. After a couple minutes all of them were all able to regain their thoughts and composure. They were able to remember their original plan and really start to plan and sequence their journey.

Just then Kira said, "Just because we made it onto the space highway doesn't mean we're safe in fact we're now in more danger than ever! Cause of the horrible crashes that happen all the time, so don't drop your guard!"

Before she was done saying that a huge cargo freighter up ahead lost control and violently started crashing into every craft in back of it. Then Abran, thinking fast maneuvered their ship in between the giant pieces of shrapnel coming towards them! As the freighter spun out of control and struck another freighter behind it! Then the two of them blew up with a giant explosion! The brothers ship passed between them just in time! Hopefully all the other crafts in back of them had their shields on or they'll most likely die in the accident too.

Narrator:

That's one thing about the space highway, even though you're traveling far beyond the speed of light you're actually being propelled by space itself. When it's referred to as the speed of space, it means the expansive movement of space/time from all matter in the universe. Space is infinitely smaller than anything else in the universe and has absolutely no mass. It travels infinitely faster than light and carries all other matter on it, almost like a river. When a particle of matter gets broken down small enough, to the size of photons or other electromagnetic particles it becomes aloft carried by space, and the smaller it becomes the faster

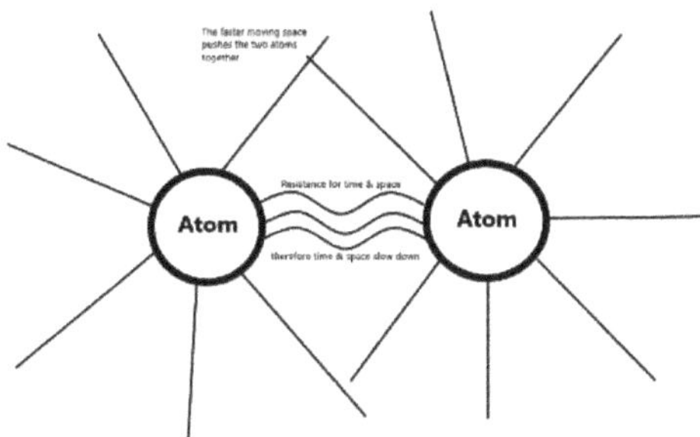

space can propel it. The movement of space creates gravity. On the Nano scale matter behaves like this illustration.

The space/time between the two atoms moves slower because it meets resistance. The space/time on the outside moves faster because it has less resistance. According to Newton's law, for every action is an equal and opposite reaction. The faster moving space pushes the two atoms together, but gravity is very weak; the only thing gravity has a lot of is time. Matter is essentially concentrated space, of course the only question is a paradox whether matter is releasing space or is the expansion pulling the space out of matter? Since space/time has absolutely no mass and light does, light cannot travel as fast as space. Therefore, the space highway is basically the focused movement of space/time essentially a network of worm holes connecting all the known universe together.

I'm sorry to pause the story to explain the dynamic functionality of the space highway, but at this point in the story it seemed necessary to do so.

We now re-join the brothers……

The brothers breathed a sly of relief. Erik almost shouted, "Damn that was Hella close, we definitely need our shields on now!" Erik immediately turned the shields on, then they all felt a lot safer. As they continued along their journey, no more major mishaps took place. Kira admitted that she had never traveled further than her home planet of Marcus 12, and really didn't know of any good destinations. So, Abran basically just randomly put his finger on the screen, his finger landed on the Reptiless galaxy which happens to be the closest one with 6 solar systems. One of the planets had a cool sounding name (Nameka) Scott said ,"Let's go there, according to the map we can get supplies there." At this point Kira knew as much about where they were going as they did. Just a few hours later they were coming up to one of the space highway junctions. Which they had already selected the direction that they were going to take.

The junctions are one of the only places where the space highway actually slows down to the point where you can take the route that you have chosen, because if you take the wrong one you won't be able to turn around or exit for millions of light years. That's why it's important to take the right one the first time. Since the brothers had already chosen the direction that they were gonna take, their slowdown time was minimal. Once they took their new path they were speeding off in the new direction. So, now there was only a couple of hours left until they'll reach their destination. During this time Scott was busy thinking [Now what do we do? I mean this adventure is all we ever dreamed about…right? It's like a dream that you thought about but never imagined that it would actually happen, or it's like a good adventure movie or a great book. Not to mention that we're fugitives and if we're captured, who knows what the consequences for our crimes would be?]

Kira was starting to realize that it has been a long time since she felt like she belonged, and the brothers had accepted her like family. Kira never knew her birth parents and was never able to find out what happened to them. She was raised in a foster home with other children of all ages without parents.

Everyone has some part of their life that has tragedy and triumph. Throughout her childhood Kira was teased from time to time about not having parents and a family of her own. After being teased she would cry and run home to her only private place. Which was a quiet meadow with tall green grass and a small freshwater stream flowing through it.

It was not very far away from where she lived. There she could be alone and daydream about the future as well as other things.

She heard that joining the military would be her ticket to freedom. Once she turned 18, she was old enough to join the military and that's exactly what she did. She felt like her military unit was a family to her and she felt happy to be part of this family. Like all good things it was a lot of hard work, but she didn't mind. Through hard work and dedication she advanced in the military ranks rather quickly to Sergeant of the radiological surveillance unit. Which was the position that she held up until the day that the brothers abducted her. Kira started to realize that she kind of let herself be abducted, because she was well trained in self-defense and could have escaped by fighting her way out. Instead she pretty much just went along with everything.

After another hour had gone by. The proximity alert sounded, which signified that they were approaching the off ramp for their destination which was the planet Nameka in the Reptiless galaxy. As they prepared for their departure, the brothers didn't know what to expect at the moment they reach the ejection point. While looking at the read out on the computer Erik started the countdown 10,9,8,7,6,5,4,3,2,1. Then...... Much to their surprise and delight the transition was fairly seamless, there was no change except mostly in the state of mind rather than anything physical. As they emerged from the off ramp with the thousands of other spacecrafts coming and going, thank goodness they still had their force shields on because once again a small freighter crashed into a couple other ships right in front of them creating a huge fireball and they couldn't avoid the crash, so they had to fly through it. Besides having to squint or close their eyes, not much else happened. They emerged from the fireball unscathed!

As all the other crafts began to disburse in different directions. In the distance they saw the planet that they were going to. The planet Nameka is about the same size as Earth and has an overall greenish tint. It has a single star about the same size as our sun and is orbited by a single moon. As they drew closer to it, they could see a small amount of spacecraft coming and going. They finally got close enough to enter the outer most atmosphere and kept descending. Erik conducted an analysis of the atmosphere which had many more layers than the Earth. After the analysis was complete, the air was found to be breathable.

Then Kira stated ,"All of the planets that the space highway goes directly to. Have undergone some sort of oxygenation processes. The reason for this is because the vast majority of intelligent life has to breath air with at least 17%-25% oxygen levels. Therefore, pretty much all of the inhabited planets, asteroids, and celestial bodies that the space highway goes directly to will contain breathable air, but it still doesn't hurt to double check."

As they sank deeper into the atmosphere, they saw some sort of outpost in front of them about 5 miles away. It was floating about 40 miles above the planet's surface. When Abran saw it, he said, "Hey, let's go there and check it out!" They all seemed to agree to go there, as they got closer to the outpost it appeared to be a saloon/outpost designed just for travelers like themselves. So, the likely hood that this place would have most of the supplies that they would need was pretty good. The outpost/saloon was about 6 acres in size. Only smaller craft were able to land directly on the outpost where docking was provided. Large freighter ships would have to park off to the side or above. As the brothers pulled up to land, suddenly out of nowhere a small craft quickly swooped in below them landing in the spot that they were going to land! Abran stopped really fast shouting, "You Idiot! I can't believe that jerk!" Erik said, "That was really rude!" Then Scott said, "Let's hope that everybody here isn't rude like that guy!"

So, Abran lifted the ship back up to look for another spot towards the edge of the lot. He found a spot that just opened up and started to gently pull in nice & slow. However, once again out of nowhere a small craft darted in and tried to steal their spot out from under them. So, thinking fast Abran had to drop the ship down really fast to stop the small ship from stealing their spot. The small ship stopped suddenly, and the pilot gave them a nasty look and the middle finger, then he darted off. Abran said what all of them were thinking which was ,"Maybe it isn't such a good idea to come here after all?" Then Scott said, "Well anyways, we're here. This place seems *kinda* ok, besides we're not gonna stay long, just long enough to get the supplies we need." Erik pushed the button to open the door. While the door was opening Scott was thinking [Not every creature out in deep space and across the universe is gonna be humanoid. They might be alien or something even stranger] All three of the brothers began to do some real soul searching as the door opened.

Abran, like usual was the most apprehensive about the situation, but he usually was the least trusting of the brothers, because Abran has been betrayed in his life by many of his so-called close friends. Sometimes Abran would be reclusive even to his own family. Erik on the other hand was a little more trusting than Abran was.

As of now Erik's main concern was about law enforcement tracking them here, for the fact that they were now considered fugitives. Scott's main concern was the safety of his crew. Throughout their short journey Kira had become an unspoken crew member.

And lastly Kira was still concerned about her close friends, they would be worried about her, and probably already started to search for her. These close friends of Kira's was her military family, and her two very best friends were Bradley and Nicole. These two people are the ones that she'd give her life for, and they'd do the same for her! That was her family, Kira knows that they are not blood relatives but nevertheless she loves them both more than anything. Kira started to shed a tear thinking about the fact that she may never see them again. The one thing that Kira could appreciate was that she was starting to feel very close to the brothers.

Feeling close, being loved, and needed. Feeling like you belong are important to all people. Which gives you the passion and drive to keep living and fighting. Having the zest for life is important because life can be unfair and downright cruel. Sometimes you win and sometimes you lose! Life definitely has cycles of high and low points.

There's definitely a feeling amongst the brothers that things might be starting an upward turn.

Once the door opened all the way. A big, hot gust of wind blew in their faces and greeted them as they walked down the ramp. Abran started to walk down the ramp first, then he stopped and turned to hand everyone a weapon including Kira. At that point Kira knew the brothers must trust her enough to give her a weapon and she trusted them too. Abran gave them the guns that they had stolen from the soldiers that they had killed on the outpost.

Abran made sure that they all knew how to use them by showing them the three power levels Stun, Kill, and Disintegrate. Then saying ,"These should disable just about anything that we might encounter." Once they were all on the ground Scott took a deep breath and said, "Well…Let's *roll!*"

Chapter 7 *"Clash of Civilizations"*

As the four of them walked towards the saloon entrance. The brothers looked up to see dozens of spacecrafts flying high above and all around them. They walked closer to the front door, there were strange looking aliens walking, crawling, and slithering all over the place. Some aliens were big, standing over 9ft. tall. Others were barely a foot tall, almost the size of elves. Abran almost started to laugh at one of the small aliens when it started yelling at one of the big 7-foot-tall aliens, speaking in a language that they didn't understand, but the small alien sure seemed mad. Erik figured that the big alien must have almost accidentally stepped on the little one.

Eventually they got to the door, and they could hear a loud ruckus coming from inside. Immediately they knew that this place must be pretty rowdy. Everything so far had reinforced the feeling that coming to this place was probably a **bad** idea.

Scott & Abran walked up to the doors which automatically slid open to reveal a very busy place.

It was kinda dark, with high ceilings and small tables about 10 ft apart. There was a mixture of unrecognizable aliens and humanoid looking creatures. They walked inside and went to select a table that was empty and looked clean. Every once in a while, a loud outburst would come from the other side of the room, it sounded like there was a gambling game over there. Scott said, "I don't know how to pay for anything, but we need stuff." After Scott finished saying that, Kira said, "I can help you pay for that. The main reason why I need to pay is because they'll use a thumb print reader for payment and there's almost certainly an APB out for your arrest. So, if you identify that will alert authorities of your whereabouts." Scott replied ,"That makes sense, anyways me and Kira will go ask the bartender if he knows where we can get supplies." Scott and Kira stood up and walked up to the counter to talk to the bar tender. The bar tender spoke to them in a language that Scott did not understand, but Kira did. Then she spoke back to the bar tender in plain Earth English. When Scott heard this, he was a little perplexed about their conversation. Kira asked Scott what to order and Scott said, "Do they have 4 cups of plain old drinking water." Kira told the bar tender what Scott said. After the bartender spoke to Kira she told Scott, "He said for us to go back to our table and a waitress will bring us our drinks in a minute or two." Then Kira placed her right thumb on the thumb print reader to pay for the water. Then as they were walking back to the table Scott asked Kira, "I didn't know you spoke whatever language that was."

Kira said, "I don't speak that language. When we get back to the table, I'll show you guys something." They got back to the table and sat down, and Kira proceeded to pull out what looked like an ear plug out of her left ear. Then she said, "I almost forgot that you guys don't have one of these. It's a universal language translator. All three of you are gonna need one of these. This is a basic model, but it still translates a lot of languages. There are some models that you implant in your brain and other models that claim to recognize all languages no matter what."

Erik was thinking of how he was constantly impressed by Kira's Knowledge about this newfound world for them, but for Kira all this new stuff was common knowledge for her. Erik was starting to have feelings for Kira possibly starting to have a crush on her. He wasn't sure about anything; he just knew that what he felt was special for her.

Just then the waitress came walking up to the table and said, "Here's your water!" and she placed it on the table, I mean she pretty much dropped the tray on the table. Luckily, the cups of water didn't spill very much.

As the four of them sipped on their water. Scott started discussing their next move and where to go. Just then out of the crowded room they heard a voice shout out, "**Kira!** Oh my god. **Kira Xena Jones** is that really you? It's so good to see you!" They all turned to see who was talking and saw a young man about 5'6" tall, with blonde hair, and blue eyes walking towards them with a big smile. The young man approached them close enough to offer Kira a hug. So, Kira immediately recognized him. She stood up without any hesitation and hugged him. They embraced for at least a full 10 seconds.

It was obvious to the brothers that Kira knew this stranger quite well. Once Kira and the blonde haired stranger broke from their embrace, Kira said, "Man it's good to see you, Bradley." Then Bradley said, "It's good to see you too Kira." Then Kira said, "Oh yeah. I want to introduce you to my friends; this is Scott, Abran & Erik." Then Kira introduced Bradley to the brothers saying, " Guys this is Bradley he's my brother well, not really my brother, but he's my best friend from way back in military school we're like family. I trust him with my life!" Bradley said, "Trust me when I say it's nice meeting you, and any friend of Kira's is a friend of mine too." Then Bradley turned to Kira and asked her, "So Kira, you're a long way from home aren't you? I remember you were stationed on that outpost on the edge of the milky way galaxy. I thought you'd be there forever, anyways where are you going? I mean, cause no one stays here for long if they can help it. As for me, I'm heading to Aqualaysha it's in the Neptorian galaxy on the other side of this quadrant." Kira said, "Aqualaysha? I've never heard of that planet."

Bradley said, "Oh yeah I forgot you haven't been that far out in the quadrant before. Well as far as Aqualaysha goes it's a water planet except for the planet's core that's where all the civilization lives for the most part. Anyways let's all sit down and get to know each other a little better and catch up on old times."

Meanwhile…   On the other side of the saloon there was a gambling ring going on. One of the alien's name is Quarez, a known thief and gambling cheat was playing a game of cards, and of course he was cheating.

One of the other players is another notorious criminal named Jaz, and he just bet a considerable sum of money.

Everyone playing was unaware that Quarez had previously stacked the deck in his favor. So, when it came time to call their hands. They all showed their cards and Quarez won! He was delighted that he had won and celebrated. So, he reached across the table to grab all of his winnings. That's when Jaz looked down in disgust and happened to see a card laying on the ground under Quarez's chair. He immediately shouted, **"Hey Quarez you're cheating**!" Then Jaz reached down to pick up the card, it was the king of diamonds. Jaz angrily grabbed Quarez by the scruff of his shirt and shouted, **"I don't play with cheaters!"** Then he violently shoved Quarez which caused him to stumble backwards until he crashed into the table behind them spilling everything on it. The group of aliens that were quietly sitting there immediately stood up and started yelling angrily with the intent to fight!

Meanwhile… On the other side of the saloon the brothers, Bradley & Kira all stopped their discussion and took notice of what looked like a brawl starting. Bradley already knew how bad these things can get. So, Bradley said, "Well it looks like it's just about time for us to leave folks!" As soon as he finished saying that, from out of nowhere a chair went flying overhead! The chair flew and hit the back of the biggest alien in the saloon! He got mad and walked over to one of the biggest tables and picked it up! Then he threw it at the group of brawling aliens! As the violence escalated, before long everybody in the place was either fighting or throwing something!

As the brothers, Bradley & Kira were trying to sneak out bigger & bigger objects went flying overhead. They had to crawl all the way to the front door.

Just after they exited the saloon, they saw the entire bar tenders counter go flying overhead and smash through the wall! Finally, the brothers, Bradley & Kira got outside. They started running towards their ship! There was a constant flurry of debris hurling out of the saloon, it was utter mayhem! Suddenly they heard a very loud crash! They all turned around to look and saw the entire building collapse! Scott said loudly, "Man that turned ugly real fast!" Bradley replied, "I know these things almost always turn out that way. Don't worry they'll have this place rebuilt in a couple of days."

After hearing that Abran said, "No way, a couple days that can't be true." Bradley continued saying, "Yeah those constructo-bots are very efficient and cheap. So, it's no big deal when that place gets destroyed, it happens about every month or so. Anyways let's get back to your ship."

Once they got to the door of the ship Erik typed in the code keys. The door ramp slowly opened, and they all went inside to sit down. Abran started the engines, and the ship slowly began to rise. Just before Abran was ready to take off. Scott said, "Wait! we forgot about the supplies! That's the reason why we came to this damn place!" Then Bradley said, "I hate to interrupt, but what types of supplies do you need?" Scott answered, "The basic stuff for traveling you know like food, drinks, hygienics." Then Bradley said, "The reason why I'm asking is because it sounds like everything you need is down on the planet's surface." Just as they were getting ready to leave Kira spoke up saying "Wait! Bradley, what about your ship don't you want to go get it?"

Bradley's facial expression abruptly changed in response to Kira's question. Bradley replied saying, "Well you see Kira the fact is that I lost my ship in a bet against that low life Quarez." Kira snarled angrily saying "Dammit! Bradley, you had a great ship, how could you let him trick you like that?" Bradley replied, "Well, I had been drinking a little bit." Kira answered, "Ahhh Dammit! you're such an idiot when you're drunk!" Then Bradley said, "Well Kira, if it makes you feel any better, it's been over a year since I've drank even one drop of alcohol and never felt better or looked back." Kira replied, "You know it's hard to believe you." Bradley said, "I know ,but I do have a plan to get my ship back because I know Quarez cheated, I just can't prove it!" Kira said, "Of course he cheated, he always cheats, you should've known that, and you would've if you hadn't been drinking." Bradley replied, "I know, I know, I'm sorry!"

Scott interjected, "Hey you two, I'm sorry to interrupt but we're getting close to the planet's surface, and we need to know exactly where to go." Then Bradley pointed out the window saying, "Right there, you see that city down there that's where we can get the supplies that you need." Abran took the ship right towards the city. As they got closer Abran found it more difficult to steer the ship through the traffic. Just then a voice came over the radio speaking in a language that the brothers couldn't understand. So, naturally they looked to Kira because she'd be able to translate it for them.

Erik exclaimed, "Well, what did they say?" Kira said, "They said that we're entering on the wrong approach vector, in order to fix that we must enter from the other side." Abran was the one who knew everything about piloting our ship.

So, he immediately changed course to enter from the other side. The radio voice started speaking again. So, once again Kira had to translate it saying, "They said we could land on platform 327" Erik said, "*Oh boy*, I kinda have a bad feeling about this, remember last time we were allowed to land. Let's have our weapons ready just in case." Abran gently and precisely landed the ship on the designated platform. Once they landed, they just waited for a few minutes. They wanted to make sure that no one came out to arrest them or something like that. Bradley said, "If you're worried, I'll go first, if anybody on this planet should be worried about being arrested it's me. That should put your minds at ease."

Erik opened the door, and they all grabbed their weapons before exiting the ship. They walked down the ramp and into the docking bay. Then they saw a door that read exit. As they got to the exit door it automatically slid open to reveal a walkway.

They quickly noticed that the walkway was very busy. With lots of people and aliens crowded up & down this small path that was lined with many small shops & eateries. They also noticed the ratio of humans to aliens was about 70/30, which meant they'd fit right in. Bradley took the lead saying, "I know this area pretty good, so let's get the stuff you need and get off this planet quickly." Then Abran said, "What about fuel, we're gonna need that too." Bradley said, "We'll have to seek out a fuel vendor, don't worry those guys are almost everywhere."

At that moment Bradley saw a fuel vendor and flagged him down to talk to him. After striking a deal Bradley turned to tell his friends that the fuel vendor would meet them back at their ship.

Therefore, two of them would have to return to the ship to assist with refueling. Then Bradley said, "Me and Abran should be the ones to go back to the ship." Which would leave Scott, Erik & Kira to continue shopping for supplies. Before the friends parted Bradley said, "Hey guys there are two things that are important; one is make it quick, and two is don't trust anybody."

After that both Bradley and Abran walked away into the crowd towards the ship.

Scott said, "Well you heard him let's get moving. I just wanted to remind you Kira that we don't have any money. So, I don't know how we can repay you, but we will somehow."

As it was explained much earlier in this story that their ship had the ability to recycle and rejuvenate food, water, and air. To some degree ,but it's always better to get fresh supplies whenever possible, and with that said.

Scott, Erik & Kira came upon the first shop that was selling all the necessary supplies. So, of course they bought all they could carry and then headed back towards the ship. On the way back they passed by a small group of soldiers standing off to the side. It looked like they were questioning some alien. As they passed by the soldiers, they overheard them asking the alien if he had seen or knows of any strangers in the area. Then they showed him a grainy picture of the brothers.

The picture must have been obtained from security camera footage. Once Scott, Erik & Kira passed by the soldiers Erik said quietly, "I'm sure that we're the strangers that they're asking about!" Scott replied, "I know, lets hurry back to the ship!"

They quickened their pace while trying to be as inconspicuous as possible.

They got back to the ship just in time to see Abran & Bradley finishing refueling the ship. Bradley just finished paying the fuel vendor then the two shook hands, and then the fuel vendor climbed into his cart and said, "It was a pleasure doing business with you." And drove off.

As soon as he was gone Scott said, "That's good timing because we need to leave quickly and quietly, because there is a group of soldiers looking for us as we speak." So, after Scott finished saying that, they all quickly walked up the ramp into the ship. Then Erik closed the door, and Abran went right to the controls to start the engine. Just as Abran was going to blast off, Scott said, "No wait Abran! that might not be such a good idea, we should take off nice and slow OK. We don't need anything that will draw attention to ourselves." Abran agreed, so that's what he did. They took off slowly to blend in with all the other crafts around them. Within a few moments they were already out in space.

The space highway on-ramp wasn't far away. So, Bradley reminded them that they needed to go to Aqualaysha in the Neptorian galaxy. Erik started thinking [Why does Bradley want to go to that planet so badly? What's so special about that planet? Maybe this is some sort of trick?] As they approached the on-ramp to the space highway.

Erik was looking at the super quadrant 368 download map and calculated there was approximately 11.6 hours of travel time from here to Aqualaysha which is 11.5 million light years away.

As they got closer to the on-ramp to the space highway, Erik said, "I'll turn on the shields now, I remember how it was last time." With the shields now on they entered into the space highway on-ramp to start the next phase of their journey.

Chapter 8 *" Further into the deepening abyss"*

After they had been traveling for about an hour everyone started to relax. Up until now everyone had been pretty quiet. The first one to break the silence was Erik saying, "Well Bradley, why do you want to go to Aqualaysha so badly?" Bradley responded, "Well, for a few reasons. Look, I can see by the way that you guys are acting that you're probably fugitives on the run. I don't know what you've done, nor do I care. Like I said if Kira calls you guys her friends then you're my friends too, and I mean it sincerely. I Know that Aqualaysha is one of the best places to hide out and lay low from the law. I have a friend there who knows a **life hacker** who can hack into the Intergalactic computer systems and reset your whole life. All aspects of your identity can be reset. They can pretty much erase your entire existence and replace it with a new one, for a price of course." Kira kinda butt in saying, "Bradley, you know that getting a life hack is a crime and the consequence for that crime is death for you and the life hacker. The crime of life hacking is considered to be one of the most serious crimes in universal law." Scott said, "It sounds like if you're successful no matter how bad you are, if you get a life hack then it's all fixed….. no wonder individuals attempt it. I wonder what the failure rate is?" Then Bradley said, "Oh yeah, I also heard a rumor that Quarez has stashed my ship there. So, I want to go there to see if it's true, and if it is true then I can steal it back, because Quarez pretty much stole it from me."

As mentioned before, that Kira had been a long-time close friends with both Bradley & Nicole. They all grew up together in a foster home on Marcus 12 in the Golden galaxy. Bradley would often talk about joining the military when he got old enough in effort to get out of the situation that they were in. They all planned on joining the military when they turned 18yo.

Since Bradley was one year older than Kira, he joined first. Then when Kira was old enough, she joined as well. Luckily, they managed to remain stationed together throughout their first year in the military. The following year they got separated, Kira was relocated to the outpost on the edge of the Milky way galaxy where she encountered the brothers.

Bradley was relocated at least a couple times after that, once to Malmactron in the Great galaxy, and then to Nameka in the Reptiless galaxy where he was last stationed.

Erik was studying the super quadrant 368 map and realized that there was two junction points and one possible stopping point along this route. He called it out saying, "We're currently on route 4211 and we're coming up on junction 3790. We need to make sure that we take route 4202 towards Aqualaysha. Once we get on that route, we'll still have 10.3 hours of travel time to go until we reach our destination."

Moments later the proximity alert sounded telling them that the junction was coming up in less than 30 seconds. Right around the 5 second mark the space highway slowed down tremendously allowing them to turn onto route 4202. Then right away they were back up to full speed whizzing off towards their destination.

Erik said, "Now we stay on this route for about 4.9 hours. Then we'll encounter another junction point and when we get to that one, we go straight through onto route 4201. That one will take us all the way to our destination."

The next eleven hours was time that allowed them to talk and get to know each other a little better. During this time, the brothers told Bradley & Kira about the fact that they are from Earth. They designed and built this ship, and the fact that they had a major confrontation with the Earth officials before they blasted off. The Earth officials were trying to capture them, and they killed their mother in the process. Also, the fact that before they entered into deep space, they had no idea that any of this stuff existed (Life/civilization), but they were amazed and delighted to find out that space was full of intelligent life.

The brothers told Bradley & Kira all of their history up to the present day. After hearing this both Bradley & Kira agreed that Earth and Marcus 12 sounded very similar. Kira said, "If you ever saw Marcus 12, you'd agree." Kira expressed sadness of how much she missed her home planet.

Before they knew it the 4.9 hours had passed, and they had almost reached the next junction 3789. Once again, the proximity alert sounded telling them that they have 30 seconds until they'll reach the junction and have to make their choice. Which was no problem because they were going straight through the junction without turning. Once again right at the 5 second mark they experienced a tremendous slow down when they went through the junction and right after that they sped up just like before.

The remaining 4 hours went by a little faster than before, and nothing worth mentioning happened during this time.

Then once again, the proximity alert sounded telling them they were 30 seconds away from the next junction point. Then… 5,4,3,2,1 and another slow down occurred to get on the galactic route 417. Unfortunately, galactic routes are slower than intergalactic ones are. With the main difference between the two is that there is more gravitational interference on galactic routes. Erik said, "As of now we only have about an hour and five minutes left until we'll reach Aqualaysha."

After the time elapsed, and once again the proximity alert sounded at the 20 second mark. They had reached Aqualaysha, Abran prepared to exit. On the 5,4,3,2,1 then… they exited and as soon as they did all the stars were visible once again and right before them was Aqualaysha, they had reached their destination!

Chapter 9 "A New Mysterious World"

They were awestruck by the profound beauty of the electric blue color of Aqualaysha. As they got closer, they saw that Bradley was right when he said there was very little land compared to water which covered over 99.89% of the planet's surface. Bradley said, "See those small islands down there, those are actually floating cities that's where patches of civilization live, but the vast majority of the population lives inside the planet's core."

Then Bradley pointed out the window at one of the small cities and said, "See that one over there, it's called Ackroneglia. I'm sure that we could buy some universal language translators there, with the exception of Kira we all could use one. So, let's land and see if we can buy some." Within a few minutes they landed. They were relieved to see that this little floating city wasn't nearly as busy as it was on the planet Nameka, especially in that rowdy saloon.

From first sight this place was much calmer and had plenty of spaces to dock, and best of all it was free docking. Before they could exit the ship Erik had to perform an air analysis along with gravity and air pressure to make sure it was safe outside. Once he found that everything was ok. So, he pushed the button to open the door. The door slowly opened, and they all stepped outside.

The second they were outside they smelt a weird fishy smell, the smell wasn't really strong, but it was bad! Right away Bradley said, "Damn, that smells horrible! It's Bad enough to gag a maggot! I almost forgot how bad this planet smells sometimes." They all walked down the ramp; Scott noticed that the waves crashing on the shore were huge compared to the waves on Earth. It was also a little windy, which helped to keep the smell down. The five of them walked closer to the stone and concrete building. There was a small sign that had words written in another language that only Bradley could read. Bradley read the sign out loud saying, "General store/ Foreign emporium."

As they walked inside, they were immediately greeted by a small flying robotic drone which came flying up and spoke in a language that only Kira could understand because of her language translator. Kira spoke back to the drone in plain Earth English saying, "We're looking to buy some universal language translators."

The drone only paused for a second, then it spoke back to her saying, "Follow me" So, they followed the robotic drone as it flew through the warehouse. Down winding little passage ways filled with various stuff stacked all the way to the ceiling. The place looked like a junk yard or a hoarder's house. When Scott saw some of the items for sale he thought [who would want some of this weird crap, most of this stuff is just plain junk!] As they continued to follow the robotic drone through the warehouse it seemed like miles of dimly lit little corridors. They felt like they were being led through the catacombs of Paris.

Scott whispered to Erik, "Why'd Bradley drag us to this dump, there's no way they'd actually have any kind of high-tech devices here." As soon as Scott finished saying that, they arrived at a door and stopped.

The door automatically opened to reveal an unbelievable room, filled with ultra-high-tech devices. They all walked in and were stunned Erik said, "I was wondering if this place had anything else besides junk." As they walked up to the counter standing in front was a retail android. The android was in the form of a beautiful human woman. She had a very soothing voice and said in plain Earth English, "Welcome to Aqualaysha's high-tech emporium, how may I help you?"

Scott was thinking that [This whole warehouse must have some type of autonomous network where the androids & drones can communicate with each other, but also think independently as well. Otherwise, how could this android already know what language we spoke?] Then Bradley spoke up saying," We're here to see the different models of universal language translators." The android spoke saying, "Most certainly, I knew that you would want to see them." Then she turned around and reached behind to pull out a drawer and then set the drawer on the counter for display. Kira selected one saying, "This one is pretty much the same as mine, it's not the most expensive. It's medium priced but it can translate about 100,000 known languages in our quadrant. I mean after all most lifeforms never leave the planet they live on, much less the galaxy or the quadrant they live in." As soon as Kira paused the android spoke saying, "This new model can translate up to 500,000 languages and is expandable to learn the phonetics of most audible languages, therefore, it's un-limited." Bradley said, "Ok, we'll buy 4 of them." Then the android spoke saying, "Is this all together?" Bradley said, "Yes" The android said "OK, then your total is 300 Intergalactic units."

Kira said, "That's not a bad price, mine alone cost over 120 units." Bradley said, "Don't worry I'll pay." Then he waved his right hand over a scanner and a small green light appeared with a beep noise that confirmed the purchase.

Then the android handed the four devices to them. Bradley was the first one to put it in his ear, after a few seconds he said, "Ooohh…It kinda tingles a little."

Then the android said, "The unit is embedding itself into the dermis, some find this to be pleasurable, and some find it to be irritating. Nevertheless, this is a necessary function that will allow the unit to secure itself and not interfere with your hearing."

Scott was the next one who inserted it into his ear and then Erik did too, neither one of them complained. Finally, when it came to Abran of course he was the most hesitant amongst them. So, when he put it in his ear, it started embedding itself, and right away Abran complained of pain and a burning feeling. The pain only lasted for a few seconds. So, once they all had their translators installed, they were all done. Then Bradley asked, "Oh yeah, since we're done here. Before we leave, I have to ask you guys about your ship's force shields, how much have you tested them?" Scott said, "Not very much because we kinda had to leave Earth in a hurry, why do you ask?" Bradley replied, "Because in order to get down to the planet's core you absolutely must have good shields because the pressure down towards the bottom of the sea is immense. I mean we're talking up to 3,250,000psi and maybe more. So, if your shields need any improvement now! is the time and place to find out. You don't want to find out when you get all the way down to the bottom of the sea that your shields need improvement, or they simply can't handle the pressure." Erik said, "Well, like we said that we didn't get a chance to test them thoroughly. So, are you saying that we could get our shields tested here? That way we'll know whether or not they can handle the pressure at the bottom."

Bradley asked the android, "Hey Ms. Roboto, where can we get our shields tested to make sure that they'll handle the pressure at the bottom of the sea." The android answered, "You simply go back to your ship, and I'll order one of our drone bots to meet you there and help you with that."

They all started walking back to the ship and just as they were told a drone bot arrived, it announced itself saying, "Good day to you all. I am drone bot#3249687 I have been instructed to assist you in testing your craft's force shields in preparation for traveling to the bottom of the sea. The highest known pressure ever recorded down there is 3,750,000psi. There will be three separate tests performed and when comparing the results between all three it will determine if your shields are adequate or need improvement. The first requirement is that someone must be inside the ship to activate the shields and make sure that the dynamic control is on during the test."

Then the drone bot called out a large testing machine which would be performing the tests. Then the drone bot announced, "The first test is concussion, inflicting rapid powerful energy blasts. Then the second test is static high-pressure. The final test is radio energy penetration and deflection, and now we begin."

Just after the drone bot finished talking Abran went inside the ship and closed the door. Then turned the shields on along with the dynamic control. Then a small apparatus emerged from the testing machine. The drone bot put the testing apparatus up close to the shields. Once the test started it made a very loud noise, you could visibly see the shields reacting violently to the loud noise, after about 30 seconds it stopped.

Then the drone bot said, "The first test is complete, now for the second test which is the static high-pressure test." The first test apparatus retracted, then another much larger test module came from the other side and moved into position. Once the drone bot was happy with the placement of the test module, then the drone bot said, "Starting second test now." The test module started applying visible pressure to the shields in a small point, this test took 30 seconds to complete. Then the drone bot retracted the test module. Then finally the third and largest test module came from the other side of the machine and moved into position. The drone bot told everyone not to look at the intense light coming from the test module. The drone bot preferred that everyone stand behind the testing machine to protect their eyes from the bright light. When this test started it had an intensely bright light and was extremely loud. The shields reacted tremendously to this test, it almost looked like they were going to fail, after 30 seconds the test was complete. Then after another 10 seconds the drone bot analyzed and compiled all the test results.

The results compiled:

Test 1: Concussion   = rating 5 of 10   satisfactory (could use some improvement)

Test 2: Static Pressure   = rating 9 of 10   excellent

Test 3: Radio energy deflection and penetration   = rating 7 of 10   very good

The drone bot told them, "Test Conclusion is that your shields have a very good ability to handle the requirements needed for safe travel to the bottom of the sea. With these results your shields could withstand almost anything in this quadrant. The estimated maximum PSI rating is up to 4,200,000. You are now certified to travel to the bottom of the sea, have a nice trip."

With this news they went over the check list to see if they needed any other immediate supplies. If not, anything else could wait until they got to the bottom of the sea. Once they had all climbed back into the ship Abran started the engines and the ship slowly started to rise. Erik downloaded the map of the planet's core to find the best and safest route to take. Once the download was complete, he entered the coordinates into the computer and plotted their trip. As they took off and headed towards the ocean Erik turned the shields on.

With a big splash they hit the water and began their underwater odyssey. Now under water and heading straight down towards the planet's core. They could watch the pressure gauge rising a hundred PSI about every 10 seconds. Aqualaysha has a layer of water over 4000 miles deep. Unlike Earth, Aqualaysha's Gravity is 2/3 that of Earth's therefore the water exerts a little less pressure, but since the water is so deep the pressure is still immense.

The core is only 1500 miles in diameter with the walls being about 15 miles thick to withstand the pressure. As they traveled deeper and deeper into the water it got darker and darker until it was pitch black. Abran instinctively reached to turn the lights on, but Bradley said, "No! ,don't turn the lights on you can only use infrared or sonar, regular visible light would attract all the predators, because it draws them like a beacon."

Erik was looking up the information about the underwater creatures and said, "There are some amazing creatures down here most of these beasts live within the first 200 miles. Then there is a different class of creatures that live from the 200-mile mark down as far as 2500 mile range. After that, the pressure is just so immense that no creature is able to withstand it. From 2500-4000 miles deep the pressure changes the viscosity of the water to more of the thickness of syrup or even jelly. When you go down that far you'll see the graveyard of ships that didn't make it.

As our friends were traveling through the creature zone most of the creature's here glow with bioluminescence. About a quarter mile in front of them they saw a giant sea serpent swimming. Its glow wasn't very bright, but since it was so big the glow could be easily seen from far away. The sea serpent was so big it could probably swallow a blue whale in one gulp. So, it would easily swallow their ship.

As soon as they saw the sea serpent, they shut off their engines and laid low being as quiet as possible. They watched the sea serpent swim past them and into the distance. As they sat there waiting quietly Bradley whispered, "If you think that one was big, I've seen ones even bigger." When Abran heard that he shuttered a little. As soon as the sea serpent was completely out of sight Abran & Erik powered up the ship and sped off resuming their trip to the bottom of the sea. Right after they passed the 2500-mile mark there was a distinct change in the viscosity characteristics of the water. For one thing it became harder to move through the water because there was more resistance, and the clarity of the water was better because any microbial life also couldn't tolerate the super high pressure. There was no water currents and no floating debris. Any surface level storms, or weather events were too far away to effect anything at the bottom of the sea.

They finally reached the bottom, which was littered with shipwrecks and the broken skeletons of dead sea creatures. As they traveled along the bottom, the PSI fluctuated between 3,800,000-3,816,000 PSI a little higher than they expected, but nothing their shields couldn't handle. Then Bradley said, "Alright we're looking for 38° longitude and 114° latitude. That should be where the airlock doors will be located. Erik said, "Well, just like you said Bradley. The airlock doors are located precisely 6 miles straight ahead. "

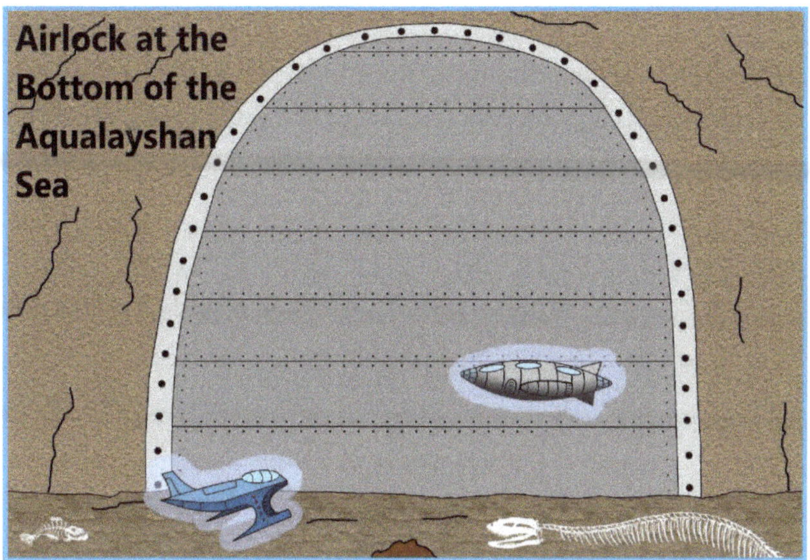

On the sonar was the faint image of an outline of a huge door embedded in the ground. As they approached the door, they detected three other ships waiting over the door. Bradley said, "Oh yeah I almost forgot to tell you that the doors only open every 4 hours or if enough traffic builds up on either side of the airlock. Especially because it takes about 45minutes to an hour to get through the doors." When they got right up to the doors they stopped. Erik said, "I wonder how long we have to wait?"

That's when Bradley said, "We should be able to tune into a data channel showing how much time is remaining until the doors open again." That's when Erik started searching for the data channel that Bradley spoke of. Then Erik said, "I found it, it says there's approximately 42 minutes until the doors open again." Kira said, "Oh great, now what the heck are we gonna do until then?" That's when Bradley said, "I guess we can get a little more acquainted with each other. Well, you guys already know a little about Me & Kira's history. So, let's hear a little more history about you guys if you don't mind. You said that you're from Earth, Well I'd like to hear a little more about this Earth that you talk about. I've never been there and frankly I've never heard about Earth until now."

Scott spoke saying, "Earth is a blue planet mostly because the surface is covered with water and the land mass is about 29%. The land is covered by a lot of green from all the plant life. There are many thousands of animal species and tens of millions of plants and insect species. Earth is a lot like Marcus 12. Until recently we didn't know that humans were so widespread across the universe. From all of our mass media they've always told everyone that there is no proof of life beyond Earth, especially intelligent life. Anyone who thinks that there is alien life must be crazy."

Kira said, "If you haven't already surmised by now that all humans didn't just originate from Earth. In fact, as far back in recorded history shows that no one is exactly sure where humans originated from. In fact, some of our greatest scientists using some of the most powerful and sophisticated super computers have calculated that all humanoid creatures will eventually form. When any planet has similar conditions like Earth & Marcus 12 which promotes the formation of life. If you look at it from a chemist's standpoint humanoid life is merely a chemical reaction. However, a complicated one."

Erik said, "I suspect that some humans have left the Earth like we did and ended up being captured and in the custody of the U.S.P. The only question is what do they do with them after they're captured."

Bradley said, "Well your guess is as good as mine, but I wouldn't doubt that they were executed. After all this universe is sadly filled with brutality and is pretty cold hearted sometimes."

Just after Bradley finished saying that Abran spoke up saying, "Hey, times up the doors should be opening right now." Right after Abran finished saying that, the ship next to them turned their lights on and the crew started moving around inside anticipating the airlock doors opening.

Chapter 10 " The World of The Turning point"

The long wait was over. They couldn't hear anything, but suddenly they saw a small beam of light emerging from the crack between the doors. They could see the giant doors slowly starting to open. They could see millions of tiny bubbles that sparkled in the light coming from the crack between the opening doors. As soon as the bubbles appeared they'd dissolve almost instantly into the super pressurized water like carbonation into soda water. As the giant ultra-heavy-duty doors slowly lumbered open they could see a bright light coming from inside. The doors took about 7 minutes to fully open. Once the doors were fully open then over a dozen crafts came rushing out and sped off towards the surface. After which a prompt came on the radio that stated, "Ok, it's all clear you now have approximately 5 minutes to enter. Any ships that haven't entered all the way cannot block the doors from closing. Any ship blocking the doors from closing will be forced to exit and wait until the next opening cycle." All the waiting ships went right in, our friends followed. Even though all the ships that were waiting had entered. There was still plenty of room for other ships to fit in the chamber. After the 5 minutes had expired some warning lights began to flash, then the doors began to close. Once the doors had completely shut, then the locking procedure started, which took an additional couple minutes. Once the first set of doors were secured, then the second set of doors began to unlock and open. The second set of doors were just as heavy duty as the first ones are and serve mostly for pressure relief, and for an added level of safety.

So, if the first set of doors should happen to suffer a malfunction there would be a failsafe in place. It took the same amount of time for the second set of doors to open. Once the doors were open, all the ships went inside the second chamber then 5 minutes later the doors started to close. As soon as the doors had completely shut and locked. Then flashing lights warned everyone that the airlock would be starting, and the water would be pumped out and would be replaced with air. As the water was being pumped out it actually went pretty fast, something over 10 million gallons of water needed to be pumped out quickly in the time of about 4-5 minutes.

Once all the water had been pumped out and replaced with air, the airlock was complete. Then both doors # 3 & 4 opened together, then all the ships were ordered to exit the airlock. As soon as all the ships were out, then a whole bunch of ships went flying in. Then the doors started closing again to do the whole procedure in reverse. At that moment Bradley celebrated saying, "All right we're in!"

As the brothers looked out the window, they saw a surprising landscape. The first thing they noticed was the good amount of artificial light bathing the land in bright even light, similar to sunlight. The other thing they noticed is how big it was, about 1470 miles across, which is not huge but was bigger than they expected.

Then Bradley said, "Oh yeah, I almost forgot the first thing they do is charge everybody an air tax, see every time they open the door a little bit of air escapes. So, therefore they must pay a courier to bring down highly compressed liquid air from the surface, which by the way is where all the air ends up anyways. You might ask why or how they found and built this underwater city at the bottom of a giant ocean under tremendous pressure. Many years ago, the Intergalactic law enforcement bureau was seeking out some planets to build one of many high security prisons throughout the known universe. The locations had to be nearly impossible to escape from, Aqualaysha was one of the planets chosen. During a mining venture they discovered that the planet's core was actually hollow. So, Aqualaysha originally was a Prison Planet. It was decommissioned about 60 years ago because too many prisoners were dying when they tried to escape. Otherwise, there was a lot of prisoners who died while just being transported here. Just getting sentenced here was pretty much a death sentence!"

Once our friends had selected a place to land and did so. Abran was the first one to say, "Well now that we're here one of the first things we need to do is to look for one of those life hackers that you talked about Bradley, so we can reset our life."

Kira said, "You know it does cost a lot, but most good life hackers will say you can just make yourself rich, but not too rich if you do that it'll cause the Intergalactic monetary bureau to look into unusually rich people, because those people might have had their life hacked. Sometimes even if they start investigating you for these things it pretty much means you're dead! So, that means if an investigator comes looking for you on suspicion of you having your life hacked, you better run for your life! The other thing is, if the Intergalactic monetary bureau decides that your assets are communal, they will just come and confiscate about 99% of your money." Erik said, "Well this is all good information, but we also need supplies depending on how long we're gonna stay on this planet."

Without feeling the need to test the outside conditions Erik started opening the door, and once it was completely opened. Just like before they all grabbed a weapon before they walked down the ramp. Since Bradley had been here before he kinda knew his way around. Even though Bradley wanted to help his friends he still planned on finding his ship and getting it back! The last time he heard that his ship was here. So, the first person that came to mind was his friend Seth. He was a local bartender working at a dive not very far from here. So, they set out on foot to find him. Bradley knew that it was a short walk from here to the bar.

After about a 20-minute walk they arrived at a place called **Roy's Hang out.** Bradley got to the door first and before he pulled the handle to go inside, he said "Man this place is old. I mean they still don't have automatic doors; this is so primitive."

They walked in and noticed how much quieter this place was compared to the saloon on planet Nameka which was rowdy with fighting and shouting. They chose a table towards the back in the corner on the darker side of the place so they could kinda hide out.

That's when Bradley stood up and said, "Hey Abran come with me let's get some drinks, I need to see if my friend still works here or not?"

At that moment Scott realized that Bradley seemed to take a liking to Abran and vice versa, not that there was anything wrong with it. In fact, Scott found this to be a good thing because Abran was always a good judge of character. When Abran either liked or disliked somebody there was good reason to take notice. So, when Abran showed that he liked Bradley & Kira that really put Scott's mind at ease.

Without any hesitation Abran got up and walked with Bradley up to the counter. Bradley spoke to the bartender who had his back turned saying, "Seth, is that you? Remember me?" The bartender turned to looked at Bradley and you could see a spark of recognition on his face, then he said, "Well, Bradley it's good to see you again! Last time I heard a rumor that you had either been arrested or you were dead. So, it's definitely good to see you. What brings you back here? Wait, I bet I know." Bradley answered, "Well two things bring me here. One is I heard a rumor that Quarez, the guy who stole my ship has hidden it here. Did you hear the same thing? If so, have you heard where he might have it stashed? And the second thing is my friends are interested in a (*whisper…life hack*)"

As soon as Seth heard that he shuttered and quickly put his finger to his mouth and shushed Bradley saying, "Man are you crazy, don't go saying that stuff here. If you want to talk about that stuff, come with me into the back room." So, Bradley turned and waved his finger towards the others signaling them to come here, when Seth saw that he got nervous. Then Bradley said, "Don't worry those guys are my close friends and I trust them with my life, anyways they're the ones who want what I just mentioned."

Seth turned around and started walking towards the back room, he opened the door and left it open. Bradley and the others followed Seth into the back room. Once they were all inside Seth demanded they close the door, so Abran closed it. Seth said, "I just wanted to go somewhere we could talk about things in a secure place, because if you're not aware there's a death penalty for life hacking! Anyways you're in luck, because I happen to know somebody who does life hacking, and I can set it up for you. As for your ship I don't know where it is, but I'm sure I know somebody who might know where it is." Erik was the first one to ask the question saying, "As far as the life hacking, how long does it take?" Seth answered, "Honestly I don't know, but I've never seen it take much longer than 3-4 hours. I Just know that it's expensive, and dangerous."

Abran asked the next question, "How extensive are the changes and when do they take effect?" Seth said, "Now wait a minute, I told you guys that I don't know. Save these questions for when you talk to my friend, he'll know everything."

As soon as Seth finished saying that he walked over to the table and picked up what looked like a smart phone and pushed a series of buttons, then someone named Craig answered. Seth told him that, "I've got some friends over who want to make some party arrangements, we'll be over in about 30 minutes. Is this OK?" Then Craig agreed and then the conversation ended, and Seth hung up. Then Seth said, "Well it's all set up, let's go." Right then Erik said, "Wait, how are we gonna pay this guy? We don't have any money." Then Seth said, "Remember that he can hack your life so that you have enough money to pay him. Well anyways he'll work out those details when we get there." Right before they left, Seth called out to the assistant bartender saying, "Hey Jeff I'm going out for a few minutes. I'll be back shortly OK" Jeff said, "Alright."

As they walked outside through the back door, both Abran and Kira said at the same time, "Wow the sky is so beautiful! It looks like the sun is starting to set" Everyone looked up at the sky to see what they were talking about. Even though the skylight was artificial it was still amazingly beautiful. It had all of the vivid beautiful colors of either sunrise or sunset, depending on what time of day it was. They all couldn't help but be amazed. After walking down, the streets and crossing alleyways, then through winding corridors, and then up & down stairways, they felt like a mouse in a maze.

After a 30-minute walk down dozens of walkways. They finally went around a corner to see the area open up to reveal a large courtyard. About 30 yards in front of them stood a beautiful Japanese style temple. As soon as they saw it, everyone let out a collective gasp from the beauty of the temple. Scott was thinking [This doesn't seem very inconspicuous for someone involved in something as illegal and dangerous as life hacking. Maybe this was a clever disguise for something else, cause it was just too obvious.] They all walked into the courtyard, and everyone took notice of the landscape that was so beautifully and meticulously manicured. Just the way you would imagine seeing a Japanese garden with every rock, stone and pebble carefully placed in simple symmetry.

The bushes and small trees were also carefully trimmed and sculptured into geometric shapes.

As the friends and Seth walked past all the beautiful scenery accompanied with the vivid backdrop of the artificial light mimicking the setting sun. The whole culmination of everything was absolutely & stunningly beautiful.

The only way to describe it was, that it was like a painting, the mere essence of a place that you would like to go but doesn't actually exist except in your mind's eye.

As they walked up to the front door Erik & Scott both started to sense that something wasn't right here. Everything about this place was just too good to be real. The brothers all looked around and started to take notice of all the subtle peculiarities of this place, somehow all the unpleasant things were missing.

Things like ;no bugs, no wind, no bad smells, no irritating noises, not even something as insignificant as dust! That's when the brothers all started to have that gut feeling. Just as Seth was gonna ring the doorbell, Scott spoke up saying, "Hey Seth, I've got a funny feeling about all this!" As soon as Scott said that Seth stopped himself from ringing the doorbell. Then he put his hand down and said, "I was hoping that you'd say that because that means that you guys are a lot smarter and more observant than the average person. As you know that life hacking is the most illegal act anyone can do, even more so than mass murder or genocide, because of what changes you can do to your life or other people's lives. Therefore, ever since you mentioned it in the bar, we started to implement a mental illusion that we telepathically implanted in your minds. So, that everything you see, hear, smell and feel is just an illusion. This technology is very proprietary, now that you are all aware of what's happening, we can discuss the business at hand. First of all, some important disclosures, once the life hack is complete, you'll need to relocate for two years. You won't be able to see most of your friends and family anymore because the risk is too dangerous that someone might squeal and that's all it takes, and we'll all be in a life threatening situation. You should know and probably do know that I'm not the person behind all this. The person or even if there is a person behind this has come across many proxy network servers to make their trail nearly untraceable and this makes it much safer for everyone involved. Otherwise, one slip-up could mean the death penalty for everyone."

Then the voice of the life hacker spoke saying, "I have researched all of your profiles. Since you're all from Earth and you guys don't have any associations outside of that planetary system. This makes your case a lot easier and safer than most. I've looked up all your records on Earth which have now all been wiped. So, if you three were to return there it would be like you never existed. I have determined that the best restart for all three of you guys is to join the United Space Police force as fresh cadette's. You'll need to go to the USP induction center located in this quadrant, in the Great Galaxy on the planet Metro-opolis. Which is located in the star group 8h882,338,881; solar system 35,727,961. Which is about 8.4 million lightyears from your current location. You will have approximately 29 hours to get yourselves to the induction center. If you don't get there and start the induction process, then you will have likely forfeited your chances at completing your life hack. Some people have tried more than once, and it almost never works the second time. Of course, when it doesn't work. It always results in the death penalty! So, you'll want to make sure that you get it right the first time. Starting right now, you three brothers have to part ways with Bradley & Kira. Don't worry you don't have to be space police forever, just long enough to complete the life hack. I also included your spacecraft in the reset. It has been given a new identification number. That way you don't have to get rid of it, but you might consider at some point getting a different spacecraft to fully conceal your old identities. When you get back to your ship, I have downloaded into your ship's computer everything that you'll need to know about your new identities so please refer to that. So, at this point I have to say that we will <u>never</u> interact with each other ever again. Therefore, I must say to all of you good luck, goodbye and GET MOVING!!!"

At that moment they heard a loud snap and then like a dream they woke up. When they did, they were back in the bar like they never left. The illusion was so real that it was impossible to discern between reality and the illusion. The mere lack of any kind of discomfort or undesirable thing was the only noticeable difference between reality and the illusion.

Chapter 11 " The New Life Begins"

Once they had all returned to reality from the illusion. They all knew that the life hack had been completed and the hacker had been paid. Now Scott, Abran and Erik knew that there was a limited amount of time to complete their life hack. They also wanted to help Bradley find his ship, because Bradley & Kira would be stranded here soon. The assistant bartender Jeff came up to talk to them saying, "Hey Bradley, Seth tells me that you're looking for your ship and I heard a rumor that it's here on Aqualaysha. One of Quarez's buddies has it. I also heard that it's not guarded by any kind of high security or even locked up. I've never seen it myself, but I'm almost a 100% sure that it's your ship. So, it sounds like it's just a matter of finding it and taking it back. They probably won't even notice that it's missing for a while and by the time they do you'll be long gone. So, anyways here's the coordinates to where your ship is. Don't blame me if it's not your ship, remember that it's just a rumor and I can't confirm it. Well anyways I wish you all the luck and I'll say goodbye and good day to you all." And with that Jeff walked off to take care of some other customers. With this new and valuable information, the friends all headed back to the ship. Once they were all inside the ship and ready to take off. The next thing that they realized was that once they found Bradley's ship. They would be parting ways and might not see each other ever again or at least for a long time. They prepared to search for Bradley's ship. Erik entered the coordinates that Jeff gave them into the computer, and it showed where the ship was. Immediately Abran & Erik started the ship's auxiliary engines, the ship started to slowly rise.

They started to coast towards their destination. They tried to remain inconspicuous by traveling kinda slow. The map coordinates showed that Bradley's ship was to the far west of a prison compound.

Abran instinctively went to turn on the cloaking shields only to see a fault saying {insufficient power for desired operation!} Then Abran remembered that the cloaking shields can't work without the main engines being engaged. Therefore, the only choice was to fly around to the other side. Then they would perhaps have to wait for the perfect opportunity to remain unseen.

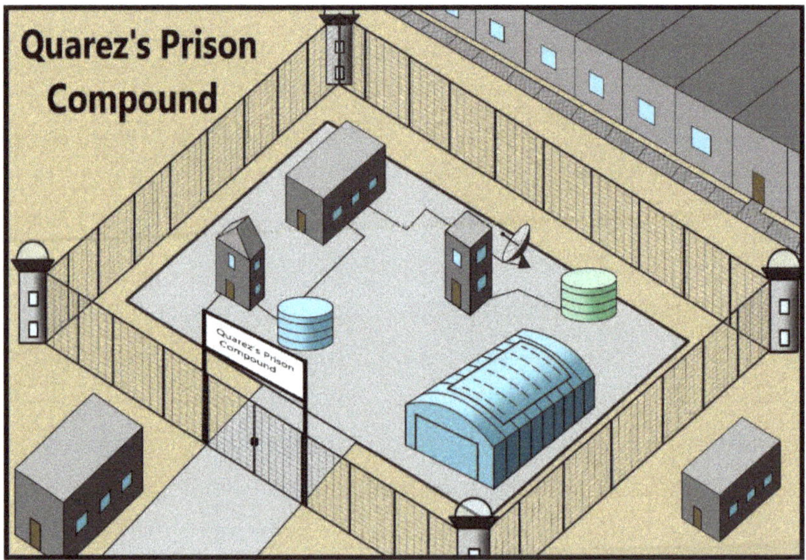

They followed the coordinates to the vicinity where the prison compound was. Then they flew around to the other side where the hangar was. With hopes that Bradley's ship was being stored inside. They tried to fly the ship as low as possible to just barely clear the tops of the surrounding buildings trying to stay below the radar or any other scanning devices.

They flew over the security fence and got right next to the hangar where Bradley's ship was supposed to be.

They lowered their ship to hover just a few feet above the ground and opened the door ramp so that Bradley & Kira could jump out safely. Before they were ready to jump, Scott handed Bradley and Kira both a weapon. Then Everyone had to say their teary-eyed goodbyes to each other. Then they all hugged and said how much they'll miss each other and the fact that once they've all departed, they will probably never see each other again. Mainly because it would be too risky to keep in contact with each other, at least for a 2 year period, or maybe longer.

Then Bradley and Kira both walked down to the edge of the ramp, and they both looked back. Erik said, "We won't leave until you guys have not only found your ship but also got it running. Then when you're ready we'll all leave together, OK." Kira said, "Ok, thanks you guys for everything. We'll miss you so much, GOODBYE!" Then Bradley & Kira both jumped to the ground and ran into the hangar. As they went through the door, the brothers couldn't see them anymore. Abran switched the display to the thermal-infrared image to see them inside the building. They could see them running through the hangar, then they stopped in front of this large dark object, which must be Bradley's ship. Through the thermal-infrared image, they couldn't see the signature of any other life forms. Unless the ship was being guarded by some sort of autonomous android or a simple robot designed to stop intruders. The brothers stood by anxiously waiting for Bradley & Kira to get inside their ship and start it. Just then Erik said, "I just detected a burst of energy, which I'm sure is their engines starting, and that's great news!" Then the giant hangar doors started slowly opening.

Scott made the comment, "I guess the rumor that Jeff heard was right about this place and it being unguarded, so far this has been too easy." Once the hangar doors were open Bradley's ship slowly flew out and got next to the brother's ship. Scott said on the radio, "Well now we'd better head for the exit!" Bradley exclaimed, "Wait, there's one problem: all crafts must register when leaving. Any craft that exits has to match the ones that have entered. If there is a mismatch it might mean that something illegal is happening and that will flag the authorities. Let's go back to Roy's Hangout to ask for some help and advice on exactly how to exit safely, because we definitely need help to make sure that we get this right!"

Once they all got back to the bar and sought out the help from Seth & Jeff. Once again Bradley spoke to Seth saying, "Hey Seth, can we speak to you in private?" Seth said, "Alright, let's go to the back room." Once they all got inside and closed the door Seth said, "What seems to be the problem? Although I think I kinda know." Then Bradley said, "Well, as you can see, I found my ship and took it back. Now we're trying to figure out how to get out of here without triggering any kind of response from the authorities." Seth said, "Well it's obvious that Bradley's ship is gonna have to use the freighter airlock to exit. The main thing you need is good cloaking shields which will allow you to sneak in when the freighters enter the chamber without being seen. Once you've entered the chamber, you have to get between the last two freighters because when the robotic drones scan the area with their sensitive optics they're less likely to detect your presence. Then you're about halfway out of the woods at that point. Then when you enter the next chamber they'll flood the chamber with water and measure the volume of water going in against the volume of air going out. It's not very accurate because it's designed to measure huge freighters, and not small craft like yours. If you release a volume of air that is equal to your ship and remain very still during this measurement you shouldn't be detected. Every 72 hours the officials take an air mass volume measurement of the entire planet's core to see how much air has been lost or gained, it has to match the other measurements. It's crucial to have the air pressurized to about 60psi which helps to counter act against the outside water pressure, which is immense. That's why the air mass volume is frequently and carefully monitored, so it can be maintained for safety reasons."

The brothers were amazed from the knowledge that Seth had about the essential infrastructure of the planet's core. Scott was curious and asked, "Seth, how do you know so much about all that stuff?" Seth replied, "I use to work for the barometrical monitoring agency for about half a year, til I was let go." Erik said, "Why were you let go? If you don't mind me asking?" Seth replied, "Because they discovered that my credentials didn't match their requirements. I don't know why they didn't discover that sooner before they hired me? Because you have to undergo an extensive background check before they'll hire you." Then Abran spoke saying, "Well good for us that you have a lot of useful knowledge." Seth replied, "Well before you leave, I have a huge favor to ask of you guys. Which is, can you take me with you and stop at the Great Salt Asteroid along the way. After hearing where you're going this stop wouldn't be out of your way. I have saved some compressed air to displace the air volume that my body occupies, and I can help you guys escape from here. Because timing is very important right now especially after a life hack, because as you know this planet is notorious for that. Any craft trying to leave this planet at the wrong time or place runs the risk of being tracked and monitored. If that happens there's a 50% chance that things won't go well for you, and you already know what happens when you're caught! So, when you're ready to leave the next time window is in about three hours and forty-five minutes from now. That's when the next shipment of surface air is supposed to be here. That's when your best chance that any air mass discrepancy would be overlooked. So, if it's settled that's when we'll go for it."

Scott said, "After hearing your case you've convinced me to allow you to come with us. We'll take you to the Great Salt Asteroid as long as this deviation won't take too long, because remember that we have time constraints. So, quickly you'd better go home and get your personal belongings and meet us back here with plenty of time to spare."

After that Seth briskly walked out to go home and get his stuff. After a little more than an hour later Seth returned with a small cart which contained his personal belongings. The first thing Seth said, "I hope this isn't too much stuff. I tried to bring just the things that I couldn't live without." Erik looked the cart over and said, "I think it's just barely under of what could be allowed. So, it should be OK, but Abran will have to be the judge of that. Well, if you're ready then it's time to go back to the ship."

When they got back to the ship Abran was already inside getting things ready to leave. He saw them outside and opened the door. Once everyone was inside Abran saw Seth's cart and commented saying, "I don't know, that's kind of a lot of stuff." Seth said, "I know, but I don't plan on ever coming back here again. So, I brought everything that's just too important to leave behind.

Don't worry I already released the volume of compressed air that I saved to compensate for me & my stuff." Abran said, "That's good to know, and as far as your stuff goes. I think we can fit everything." Seth said, "Thanks you guys, you don't know what this means to me. I've been searching for a way to get off this planet and now it's finally happening. I don't ever remember being happier!" Then Erik said, "I'm glad we could help you. Well anyways, we got to get going you might want to sit down and get comfortable." Just then Erik called on the radio to Bradley & Kira and said, "Bradley do you copy? We're leaving now, how 'bout you guys" Bradley answered saying, "Yeah we're leaving too." Scott said, "You two better take care of yourselves and trust me, we will see each other again someday and thanks for everything. I mean that with full sincerity. We will try to contact you guys at least one more time when we get to the surface. After that we can't make any other contact with each other again, at least for a while. Even though our transmissions are encrypted, we just need to play it safe." Bradley said, "I copy that, and I wanna thank you guys for everything. If it weren't for you guys, I would have never been reunited with Kira and got my ship back. Now that I have my ship, me & Kira have decided to go to Marcus 12 to see if we can catch up with our old friends Chris & Nicole. The last time we heard that they are still there. So that's where we'll start looking for them first. Anyways, I want to say thank you guys for everything and goodbye!"

Then the radio went silent, Bradley & Kira sped off in the opposite direction towards the freighter airlock. Right before they disappeared over the horizon the brothers saw a rhythmic flash of light coming from Bradley's ship that looked like they were signaling goodbye, then they vanished.

There was a moment of silence. The brothers all felt an intense wave of sadness come over them. After that moment of clarity, the next thought was strictly about the journey ahead.

Scott was thinking [Well amazingly we had made it this far. So, even though the path ahead is quite scary, accompanied with the fact that the pathway back did not exist anymore.]

Scott snapped out of his daydream to Erik yelling out, "Hey Scott, we've got to go right now! It's the window of time that we've been waiting for!" Then Abran slowly engaged the auxiliary engines and they headed towards the same airlock that they entered through. They got to the airlock doors just as they were starting to open. The entry light was red which meant that there would be craft exiting. As soon as the doors were fully open around 30 small craft came flying out. Then the light turned green, and they raced into the airlock along with a whole bunch of other craft.

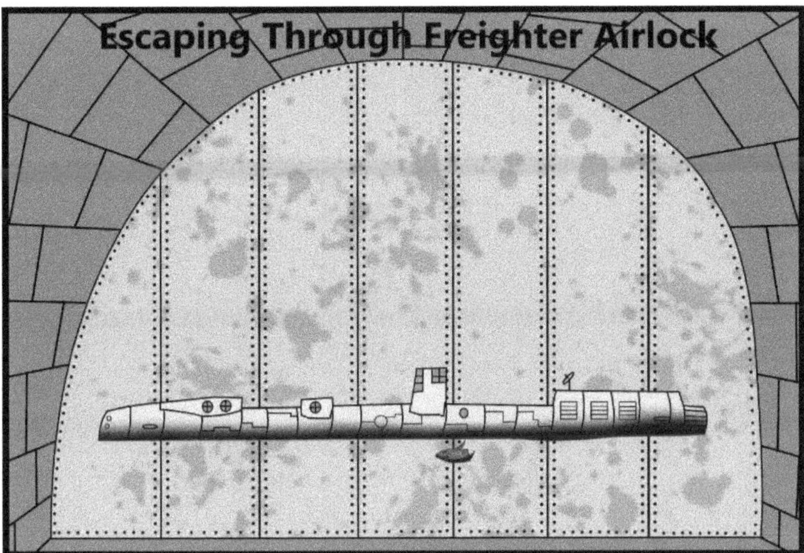

Meanwhile… On the other side of the planet's core Bradley & Kira were embarking on a much more dangerous escape. Which would mean that if they were caught it would probably mean prison time or even a death sentence! As Bradley & Kira got closer to the freighter airlock Kira turned on the cloaking shields, just as this huge freighter came towards them. They stayed very close to the freighter as it started to enter the airlock. The freighter airlock was at least 5 times bigger than the other one was. Once the freighter had fully entered into the airlock and was trying to straighten out. Bradley & Kira had to remain dangerously close to the big freighter. Remembering what Seth told them, that they had to stay close and between the last two freighters. Once they had positioned themselves next to the first freighter. It was very nerve racking because they couldn't move while the other freighters got dangerously close. If they moved, they might give themselves away, the only thing they could do was sit there motionless and hope that they wouldn't be crushed or discovered.

After an agonizing 30-minute wait for all the freighters to be fully in position. Then a small door next to the control room opened and two small robotic drones flew out and began searching & scanning around the chamber for any stowaways. Bradley & Kira pretty much held their breath. Bradley whispered, "This search is one of the most important ones because the drones have very sensitive optics and sometimes they can pick up even really good cloaking shields. The slightest movement is usually what gives you away." Just when Bradley finished saying that one of the drones came in their direction from the front of the freighter then it stopped and turned towards their ship. The drone started scanning. Kira almost started to cry thinking that they would be discovered, Bradley shushed her. They waited quietly for a VERY intense 15 seconds of scanning. When the drone stopped scanning, it obviously found nothing and moved on. After both drones had finished their scan of the entire chamber then they went back into the small door they came out of. Then shortly after that the doors closed, and the airlock started.

Both Bradley & Kira breathed a major sly of relief! At this point they felt like they were pretty much home free.

Then the doors to the second chamber started opening. When they had completely opened, all of the freighters slowly moved inside. After the doors closed and locked. Then from a giant drain opening in the floor a huge amount of water started flooding into the chamber. Bradley said, "At this phase they will be measuring the volume of water coming in against the predetermined volume of air going out and there shouldn't be any discrepancies at this time. Right now, I'm gonna have to lower our shields because I brought a canister of highly compressed air that we can release, and it should match the volume of air that our ship displaces. We will need to release the air right before they begin this process. If we're successful at matching the volume of air, to fool their instruments then we will truly be home free at that point."

Bradley lowered the shields and activated the computer so it could release the precise amount of air. Just as the computer finished releasing the air. The water had already filled to about halfway from the top of the chamber. Bradley said, "There, that should be enough to equal our displacement, I guess we'll know for sure if the next airlock opens or not."

Just after Bradley finished saying that the next set of doors started opening. Kira & Bradley both breathed a huge sly of relief and whispered joyfully, "We made it!" Then Bradley said, "O.K. only one more airlock to go."

As soon as the doors to the last chamber fully opened all the freighters moved inside and waited for the doors to close. Then Bradley said, "Right before the main doors open to the outside, we're gonna have to turn our cloaking shields off and turn the force shields on. When the main doors open there will be a major pressure increase. We'll need our force shields to be at full power. Right now, I need to move our ship further away from the freighters before they turn their shields on so that we don't get caught up in their shields. Once we turn the cloaking shields off there is a chance that they'll see us, but at this point we don't have much of a choice."

Just after Bradley finished saying that, the freighters all turned their shields on which meant that there is only seconds until the main doors will be opening. Bradley & Kira were exhilarated with the tantalizing prospect of freedom at hand. Even if they were successful at escaping, which looked very likely at this moment. They still had a long way to go until they reached their destination. Just then they could see the doors starting to unlock and open then Bradley said, "Brace yourself!" As soon as there was a small crack between the doors they felt the pressure shock. They heard the computer give a pressure Alert! Of 3,816,000 psi which was higher pressure than usual. Bradley was a little worried about the shields holding up to the immense pressure. As soon as the doors opened enough for them to get out, they whizzed out as fast as they could to get ahead of the freighters to reduce their chances of being seen.

As they got closer to the surface the radio crackled to life with the sound of Erik's voice calling, "Bradley, Kira, do you copy? Over." Right away Bradley answered saying, "Yes, this Bradley & Kira over." They heard Scott answer back saying, "Well it's good to hear that you made it. After all your journey was more dangerous than ours was. Well anyways it's good to hear from you guys, but we have to say goodbye now. I'm pretty confident that we will meet again someday, and we can't thank you enough. We will repay you someday believe me! I don't know how, but we will. Until then thanks and goodbye."

Right after that the conversation ended and the radio went silent......

[We now rejoin the brothers]

The brothers soon realized that even though they had safely made it off Aqualaysha. The path ahead was still very dangerous. Even if they were able to take Seth to his destination. They still had to get themselves to the USP induction center and sign up as new recruits. And all this had to be done within less than 25 hours or risk having their life hack fail. The scary thing was that there was so many pitfalls ahead.

Now each move had to be carefully choreographed in order to make enough time for everything that they had planned on trying to accomplish. Along with the threat that if their life hack was to fail, there would be a lot of lives at stake. At this point the brothers needed all the luck and divine intervention they could get!

Chapter 12 *"New Places. New Friends"*

As the brothers put Aqualaysha behind them. They headed straight towards the on-ramp for the space highway. Erik had already plotted the routes that they would need to take in order to reach their two planned stops. They smoothly entered the on-ramp for the galactic space highway route 417 that crosses the Neptorian galaxy. The travel time is a little more than 3 hours until they'll reach the other junction to get on the intergalactic route 4201. Which will take them to their first stop, which is the Great Salt Asteroid.

After they had been traveling for a while, that's when Scott asked Seth, "So, Seth why don't you tell us a little bit more about yourself, if you don't mind?" Seth said, "Yeah, I don't mind. Well, my home planet is Schinické. It's in the middle of the Ghost galaxy which is located towards the edge of quadrant 366. It took me a little more than a year just to reach the closest space highway on-ramp in Super Quadrant 367. From there I was able to get to the Great Salt Asteroid. There I got a job as a delivery courier supplying goods to small communities. I met my friend Jesús on one of my routes. After some bad things happened to me and Jesús, I went to planet Aqualaysha to try to get a life hack just like you guys did. Unfortunately, I couldn't get a life hack for various reasons. The reason why I need to go back to the Great Salt Asteroid is because Jesús and I inscribed an encrypted digital code there. This code is for unlocking a fortune in legitimate Intergalactic units(money). To conceal it we hid it in an obvious public place. It's so obvious, that it kinda makes it inconspicuous. Only me and Jesús know exactly where the code is and how to decode it. So, I need to get back there because I'm afraid somebody will eventually figure it out." Scott said, "Ok, don't worry we'll get there soon enough."

Erik being inquisitive asked Seth, "So, you're from quadrant 366, it's just a quadrant and not a super quadrant, what's the difference?" Seth said," There are two main differences between quadrants and super quadrants. One is that Super Quadrants are 4 times bigger than quadrants, and two is that super quadrants usually have the space highway going through them, while quadrants either don't have any space highway traversing through them or they have very little access. That's the reason why it took me so long to get to the Great Salt Asteroid. Without having access to the space highway, space travel is very slow. There are also Ultra Quadrants which are 4 times bigger than Super Quadrants, there are not as many of those. Quadrant 366 is kinda small; it only has about 3 galaxies in it, and only one of them is inhabited. The main reason why the space highway doesn't go through it is because there is so few civilizations in that quadrant. When I think about it, I miss my home planet very much and I plan on going back there real soon, maybe after I get the money, I can buy a ship." Abran interrupted saying, "Well guys it's been around 9 hours since our life hack, which leaves us about 20 hours til we have to get to the USP induction center." As soon as Abran finished saying that, the proximity alert sounded indicating that the junction was getting close. That's when Erik said, "Make sure that we take the junction to route 4201. Then along that route we'll come across the Great Salt Asteroid."

They got to the junction point and took the right path to get on route 4201. Abran said, "Well now, we definitely have to have our shields on. At this point we're a little more than 2.7 hours away from our first stop. I hope all goes well because we don't need any delays right now." By now the brothers had become accustomed to the dangers of traveling on the space highway. Scott said, "There's no way that anyone would survive more than a minute without shields through this section of the space highway." Then Seth said, "You're right there's usually horrible accidents through this section." Abran said, "Oh no! I think something's happening up ahead!" They witnessed a small craft up ahead trying to go around this huge freighter. The small craft cut in front of the freighter causing it to over correct to avoid crashing into the small craft, but the over correction caused the freighter to slam into another freighter next to it! The collision pretty much obliterated both of them, which resulted in a giant fireball that engulfed the whole area destroying all the surrounding ships that didn't have their shields on!

The brothers had no choice, they had to go through the fire ball! They all held their breath as they went through the explosion! They had to hope that their shields would hold up as the fireball engulfed everything around them! They felt some big chunks of debris hit the shields very hard, but the shields took the brutal shock without any problems! What a relief as they passed that near disaster. They hoped that they wouldn't encounter anything worse than that. Abran said, "I think we better turn the shields up another 10%, this time seems to be a little more dangerous than the last time. We haven't seen this amount of carnage until now."

They were all stunned to see how dangerous the space highway could be. After the shields had deflected that terrible explosion they felt safer and were able to relax and settle down.

It had been about an hour since the last major accident. Seth resumed telling them about his home planet. Seth described his home planet saying, "On My home planet we don't use any money because no one needs to work anymore because pretty much all the work is done by robots, androids, or drones. So, all the people do is just create and explore artistic desires. Being creative is one thing that A.I. simply cannot do. Without needing money and all the other things that go along with money, such as killing, stealing, lying, and cheating. Combined with all the other atrocities that humankind does to each other while trying to possess other people's stuff. All this is due to jealousy and envy. Our planet right now is the closest thing to being utopia than any other planet that I know. That is if it lasts? Or how long it lasts? Our scientists have completely boiled down all biological life to a mere chemical reaction. Even intelligent life is put in this classification, provided there is proper oxygen and all the other components. Then, there needs to be enough ambient energy to start and maintain the chemical reaction. With sufficient time to let the chemical reaction run its course. Then eventually you'd end up with intelligent life. Even chemical reactions such as intelligent life forms are predictable, given the correct ambient conditions like atmospheric pressure, gravity, heat, & time. Unfortunately, the idea of everyone having <u>Free will</u> is an absolute illusion. There is a predetermination of all events based upon the total mass and energy of the known universe. The principles of cause and effect dictate everything, however it's on a much more complex level than I can describe."

Then Erik said, "I guess I can agree with that on some levels. However, after hearing about your home planet it sounds like a place that I'd like to go visit someday."

Then Erik told Seth about Earth and how their planets were very similar, but yet in their own way very unique from each other. Then Erik told Seth about their former life on Earth and why they were fugitives until they got their life hack.

"Everything effects everything else in an infinitesimal way that is unmeasurable. The unseen, unmeasurable force that drives everything can only be described as GOD there is no other word that quite fits."

By now about 2 hours had gone by then Erik said, "Well, there's only about 40 minutes until we reach the Great Salt Asteroid are you getting anxious?" Seth answered, "Heck yeah, Once I find and decipher that code then the first thing I'll buy is a ship to get back to my home planet." Abran said, "Like you said hopefully nobody has figured out that it's a code and deciphered it, but what if somebody did? Do you have a plan B?" That's when Seth said, "Well no, I really don't have a plan B, but I guess I'll figure that out when I get there."

Erik announced saying, "Hey guys there's about 25 minutes left." Right about now the brothers noticed a sharp increase in traffic to a dangerous level. The number of accidents happening was almost one every minute.

As Scott watched the other ships around them crashing and exploding, he was thinking [Man, these other ships either don't have any shields at all or their shields are just not strong enough to protect themselves, how sad they don't have to die.]

The amount of death, destruction and carnage happening was astounding. The brothers and Seth were sure glad to be safely tucked in their ship while witnessing the destruction.

After about ten minutes the number of accidents slowed way down to the point of being totally calm again for the last few minutes before they reached their destination. Finally, their exit had come so they took the exit ramp. They went from traveling at a mind-blowing speed down to just a mere couple million mph.

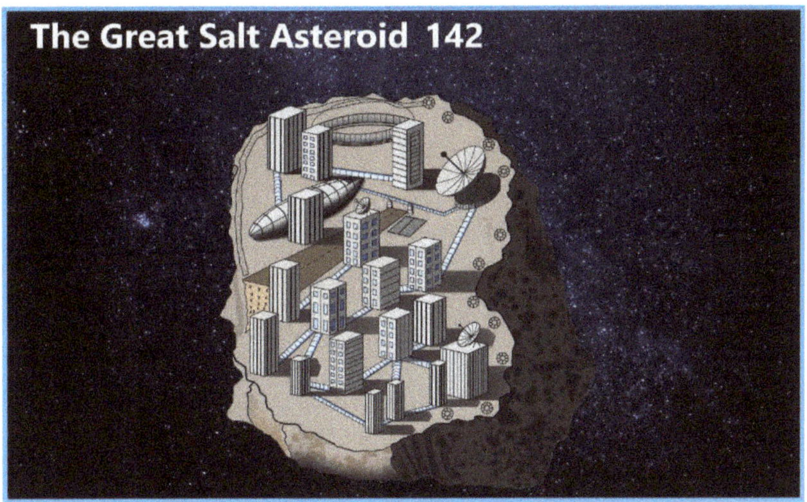

Visibly they couldn't see anything, but through their instruments they knew that they had arrived at the Great Salt Asteroid. From a distance of 5 million miles, which is the closest that the space highway would dump them. They could see that it was faintly lit up by a nearby dwarf star. The light was a little brighter than a full moon.

As they got closer, they could see lots of city lights and buildings through their front window. This place seemed to be densely populated.

Even though the amount of star light was dim the city lights made it bright enough to see all the buildings and other structures. As they got close enough to see all the space craft flying about they realized how busy this place was.

Abran was thinking [Man for such a small outpost it sure is **hell of busy**, I don't like this!] Erik's thoughts were very similar [There's so… many people here. This isn't good. Especially since we haven't finished finalizing our life hack yet.] Then Seth spoke up saying, "Hey guys can you do me a favor when we land, please don't leave until I actually found the code and try to decode it. Because once I do I'm sure I'll be OK. Otherwise I might be stranded there for a while." Then Scott said, "We'll either way, whether or not you find and decipher the code you can't come with us. We won't be able to take you anywhere else right now because of our time constraints. So, even if your plans fall through it doesn't change anything. We'll be able to come back and get you after our training. I'm sorry to say this, but I hope you understand? " Then Seth said, "Yeah I understand."

They were now close enough to receive communications from the traffic control systems. They also received a download of the docking maps, social life, laws, entertainment, monetary conversion shops, and medical care centers along with everything else that this place had to offer. Erik and Seth looked over the map for a place to dock that was close to the location where Seth thought the code might be. His friend Jesús had encrypted it near the main communications building. Seth said, "The map shows that we should request to land in the underground docking area. We could quickly gain access to the city's transportation system and reach anywhere in the city very quickly."

Once they got close enough Seth asked to speak over the radio to request for docking permission. He figured that since he had experience coming here, he would have the best luck at getting the docking space that they wanted. At that moment Erik remembered that who or whatever they would be communicating with would ask them for their ships' ID#. Right away Erik started looking for it and was surprised that even their ship's computers had been affected by the life hack, which caused the ID# to change.

Just then the radio came to life with the voice from the Space Flight Control Center saying, "Unidentified craft please state the nature of your visit and commence with the transmission of your ships ID#." Hoping for the best Seth said, "We're here on a recreational visit." Then Erik pressed the send button for the ship's ID#. They all kept their fingers crossed. Within about 20 seconds the flight control answered back saying, "Welcome visitors! To the fun, fabulous & beautiful recreational heavenly body known as the Great Salt Asteroid. You may dock at docking bay 94. You will be allowed up to 48 hours of free docking, enjoy your stay. Thank you!" Then the transmission ended. Then Seth said, "Even though we weren't given a choice on where to dock. I know that docking bay 94 is in the underground docking area, and that's exactly where we wanted to go, it's like they read our minds, let's hurry." Abran brought up the map that had been downloaded to show how to get to docking bay 94.

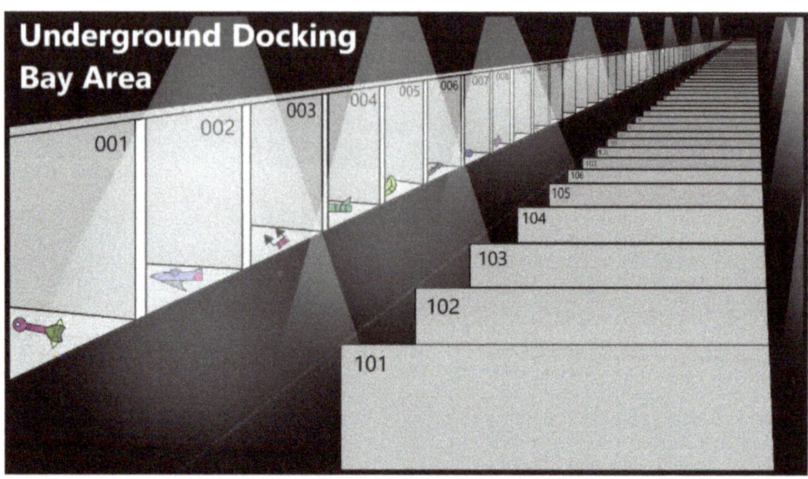

Once they got to the entrance to the underground docking area, they waited for the doors to open. After about 20 seconds the doors had fully opened. Abran carefully guided the ship right through the entrance into the underground docking area. They flew down inside and saw all of the docking bays were in numerical order and very easy to read. One thing that the brothers noticed is that since they were assigned to docking bay 94, it must be very far down towards the end of the ginormous corridor. They could see down towards the end where the lights seemed to get dimmer and dimmer the further down they looked, maybe it was an optical illusion? As they passed by the 30th bay the rest of the bays all seemed to be empty. Around the 60th bay and counting they became certain that <u>all</u> of the remaining bays were empty. At that point, the corridor took on a very eerie feeling. Seth commented saying, "Why would they assign us such a distant docking bay especially when all these ones are empty? Are they reserved or something like that?" Abran mentioned, "Even if we wanted to land in one of these empty ones, we couldn't because they all have shields blocking the entrance."

As they sped along through sections 79-90 the lighting indeed started to become noticeably dimmer. Originally Scott thought that it was an optical illusion, but clearly this wasn't the case. As they looked out the front window there appeared to be smoke floating in the air. They realized that this is impossible because there's no air outside.

Abran instinctively turned the lights on, and the smoke rapidly disappeared into one of the overhead service ports. It looked as though there was some sort of intelligence behind the smoke's behavior, and it got scared away when it saw the light. After seeing that everyone became a little un-easy. Everyone started to realize that they may have just seen some sort of apparition, which made this whole section of corridor start to feel a little haunted. At that moment they finally reached docking bay 94. They could see the shields turn off allowing them to enter and land. As soon as they landed the lights turned on and the docking bay shields went up, and everything was perfectly normal.

Erik checked the sensors, and everything was good, air and gravity were normal. Abran pushed the button to open the door, and everyone grabbed their weapons. Seth reached into his cart and pulled out a little high-tech device. Scott asked him what it was, Seth told him that it was a micro scanner, and he needed it to scan the code to decipher it. Everyone slowly walked down the ramp. As soon as they got out Seth went over to the control panel on the wall and pushed the button to call for a shuttle car.

He knew it would take about a minute for it to arrive. Abran said, "I just wanted to remind everyone that we only have about 13 hours until our rendezvous at the USP induction center."

That's when Scott turned to Seth and said, "Ok, Seth as you know we can't stay here for long, but we'll make sure that you at least find that code. After that we've got to go, I know you understand why. That gives us a little more than an hour to find your code. We'll of course go with you to help you look for it. Whether or not you find it, we'll still have to leave you here. So, you can either wait here until we come back to get you, or you'll need to figure out another way off this asteroid. Sorry to say but those are the facts, are we clear?" Seth nodded his head to show that he understood.

Shuttle cars are almost completely automated. The passengers must either type in or say the destination that they would like to go. Shuttle cars travel within a vacuumed sealed plastic tube in either a forward or backward direction. Shuttle cars all meet at the central station. From there you can take the one that will get you the closest to your destination. In order to get to different destinations, you have to change shuttle cars. There are other ways to travel around the asteroid, but shuttle cars are undoubtedly the fastest.

They heard a bumping noise followed by the sound of compressed air being released from behind the door. They all knew that the shuttle car had arrived. The door opened and Erik said, "Well let's get moving."

They all walked over to the shuttle car and climbed inside. Seth sat in front, once everyone was seated Seth pushed some buttons on the control panel telling it where to go.

Then the shuttle car door shut, and they sped off at a very high rate of speed heading for the central station. The entire asteroid was only about 30 miles wide and 50 miles long.

So, it usually takes less than 5 minutes to get anywhere. Within about a minute the shuttle car came to a stop and the doors opened. They had arrived at the central station. They looked outside and one thing that they noticed right away is the number of humans and alien life was about 50/50, so once again they'd fit right in.

The area was quite busy, Scott couldn't help but be surprised about the number of humans this far out in space. Which signified that these humans were not Earthlings they must be from somewhere else.

Abran kept reminding them about how limited they were on time. Erik said, "OK, Seth you lead the way." Seth started walking and they followed, he walked towards the main communications building which wasn't very far away, it was a little less than a mile from the central station. When they entered into the glass hallways that connected the buildings together. Erik kinda wondered [There must be some kind of shields protecting these fragile glass hallways from meteorites and other small space debris that could easily shatter the glass?] He guessed that being able to appreciate the beauty of space was worth the risk. As they walked towards their destination the brothers enjoyed the simple exterior architecture of all the buildings. Which were mostly short and rectangular shaped, but the insides were extravagantly decorated to have different themes.

After a brisk 15-minute walk they arrived at the communications building where Seth believes that the code is inscribed. They walked around a corner and Seth saw what he was looking for. A plaque on the wall representing a high-level governmental seal. Abran said, "That's it? Are you sure? Now what?"

Then Erik said, "After seeing it, why couldn't you have just wrote it down or taken a picture of it?" Then Seth replied, "Taking a picture or writing it down wouldn't work. That's why I brought this scanner with me to decode it, but I'm not entirely sure how to use it. See, I haven't told you the whole story yet. My close friend Jesús is the one who placed the code here, and he's a technological genius. He encrypted it and everything that's why I'm not sure if I'll be able to decode it without his help. He acquired the code from a friend who works in the regional governor's office and planned on using it. I was with him when he inscribed the code, but he was arrested on the false allegation of being a life hacker. So, my main reason for doing this is for him. I planned on buying a ship and helping him escape his captivity. He's being held very far away on the Prison Planet in super quadrant 92. Now after hearing this, do you understand my plight? It's not a selfish plan at all and it's the whole truth." Erik asked ,"So, Seth how come you didn't just use the code right then and there, before your friend got arrested?" Seth answered, "Because we'd have to wait for about a year to be able to use it or we could trigger some sort of investigation. At this point in time, it's been over a year."

Seth waited for the few pedestrians that were in the area to walk by and out of sight. Then when it was all clear he walked up to the inscribed code on the wall and pointed the scanner device at the code. He pushed the button, and a red laser scanned the code followed by a beep noise to confirm a successful scan. Then Scott said, "Well, is this gonna take a long time because we need to get going about now." Seth said, "I don't know how long it'll take. Because I'm not sure if I can decode it by myself. Otherwise, I'd say maybe a half an hour."

Scott replied, "Well, Seth after hearing that I'm afraid that we just can't wait that long. We're gonna have to leave you here for now. Once we've completed our mission and have become USP, we'll stop by and pick you up. Also, if you're planning to rescue your friend Jesús from the Prison Planet, I'm sure that you'll need our help. If you've managed to decode that code and buy a ship to get off this asteroid. I'm sure that you'll find a way to let us know so that we don't come back here for nothing. Otherwise, you can count on us coming back to get you, **I give you my word.** You already know about our incredible time constraints. Listen, I know that you don't know us very well but trust me, OK." And with that Seth gave each one of the brothers a firm handshake and a quick hug. Then the brothers all turned around and briskly walked in the direction of the shuttle car heading back to the ship.

Right before they rounded the corner the brothers all looked back to see Seth still standing there, they all waved to each other. Everyone knew that this could be the last time that they might ever see each other again. Then they rounded the corner and Seth was gone out of sight. Everyone felt sad, this is the second time that they had to say goodbye to a close friend.

At this point, the brothers all picked up their pace. After dodging dozens of pedestrians, they arrived at the central station. There happened to be an unoccupied shuttle car that was destined for the underground docking area. They climbed inside and Erik typed in docking bay 94 into the control panel. Then just like before the doors closed and shuttle car took off like a rocket! Heading towards their ship. Within less than a minute the shuttle car came to a stop right at docking bay 94. The door opened and they saw that the dock lights were very dim.

In the dim light Abran looked at the ship and in the corner of his eye he saw something move in back of the ship. He focused his attention there and witnessed what looked like a shadowy figure or a blurry silhouette of a person moving behind the ship. Abran instinctively knew that this was probably the same apparition that they had seen earlier. This time it took on a human form and was floating in back of the ship. Abran could barely believe his eyes, but cried out saying, "Hey guys look behind the ship there's something that looks like a ghost!" Both Scott and Erik looked behind the ship and caught a glimpse of the apparition that Abran was talking about. They all saw it right before it faded away and disappeared into thin air. Abran said, "Guys please tell me that you saw that, right?"

Scott and Erik both said, "Yeah we saw that." Then Scott said, "I don't know what the hell it was other than a ghost. You wouldn't think that there could be a ghost in space." Then Erik chimed in saying, "I don't know what we just saw, but clearly it was something that we can't explain. Maybe it was a ghost or some other optical illusion. Nevertheless, I'm still skeptical that there isn't another explanation for what we just saw."

Once the brothers got out of the shuttle car the docking bay lights detected motion and turned on. They headed straight for the ship and Erik entered the code to open the door. As soon as the door was opened the brothers ran inside. Then Erik closed the door. Abran and Scott both ran straight to the controls and started the auxiliary engines. The ship started to lift off, and the docking bay shields turned off the instant the ship left the ground. They slowly flew out of the dock and gradually picked up speed heading towards the entrance.

As they were zooming past all the docking bays they noticed that this time most of the bays now had ships docked in them.

Once they exited the underground docking area, Erik turned the shields on. Then in the distance they saw the same on-ramp for the space highway to get back on route 4201 that leads them to junction 3789. Within about 4 minutes they reached the on-ramp and smoothly entered the space highway. The brothers were getting very accustomed to traveling on the space highway.

Erik started to plot their course to the Great Galaxy, by now they had a little less than 6 hours to complete their life hack. Abran said something that was a little unsettling, which was, "What if we get to the USP induction center just to find out that the life hack was unsuccessful. Then we're captured and put to death!" That's when Erik said, "Ahhh, come on bro. We just gotta have some faith, you know cause faith is all we've got right now." Then Scott said, "Yeah Abran, I second what Erik just said."

Then Erik said, "Besides, everything that's happened so far would indicate that the life hack <u>was</u> successful." Then Abran said, "Well, you can understand why I'm worried. I mean we're going back into the heart of the enemy, into the lion's den. When we get there, they'll scrutinize our credentials closer than ever. So, if there's any detail that was over-looked or just simply missed, they'll probably find it." Then Scott said, "Well Abran believe me, I do understand where you're coming from but remember, faith brother... faith."

The proximity alert sounded telling them that they were getting close to junction 3789. They needed to take the path to get on route 4217 towards the Great Galaxy.

Taking the right path is absolutely critical at this point because they only had about 5 hours left with no margin for error. The space highway started its usual slow down as they approached the junction. Abran and Erik worked together to make sure that they took the right path at the junction. At that moment, the traffic miraculously thinned out considerably to allow them a very easy transition. Scott was thinking [Usually when things work out like this, you can say that it was meant to be. This is a very good sign!]

They were now in-route towards the Great Galaxy. With a little more than 3.8 hours until they'll get there. Erik double checked the course he plotted to make sure that it was the fastest one possible. He knew that their time was running low, they had less than 5 hours left to complete the life hack.

During the remaining 3.2 hours of their journey not much happened. While they were discussing the events at hand and what they might expect.

They were startled when the proximity alert sounded once again telling them that they had finally reached the Great Galaxy, and the next junction was coming up. They had to choose the galactic route 447, which is slower due to the fact that it is a galactic route. From there it'll take them about 1 hour to reach the solar system where the planet Metro-opolis is located.

They got to the junction and transitioned onto the last leg of their journey. That's when they started mentally preparing themselves for what to expect when they got to the Induction Center, because it was getting anxiously close.

They were now close enough to receive information downloads from the planet. Then using the downloads, they were able to pinpoint the exact location of the Induction Center. This would ensure that the moment they exited the space highway they could travel directly to their destination in a straight line without any diversions, saving precious time. Then Erik searched for images of what this place looks like. He also found some sort of propaganda video that really emphasized how great this facility was. As they watched the video, they saw that this facility looked like it was very high security. Therefore, if the life hack really needed to work, now was the time.

From the video, they saw that every recruit who wants to join has to undergo an extensive background check before anything can happen. That part alone is the one that made the brothers a little worried.

Chapter 13 "The Induction Begins"

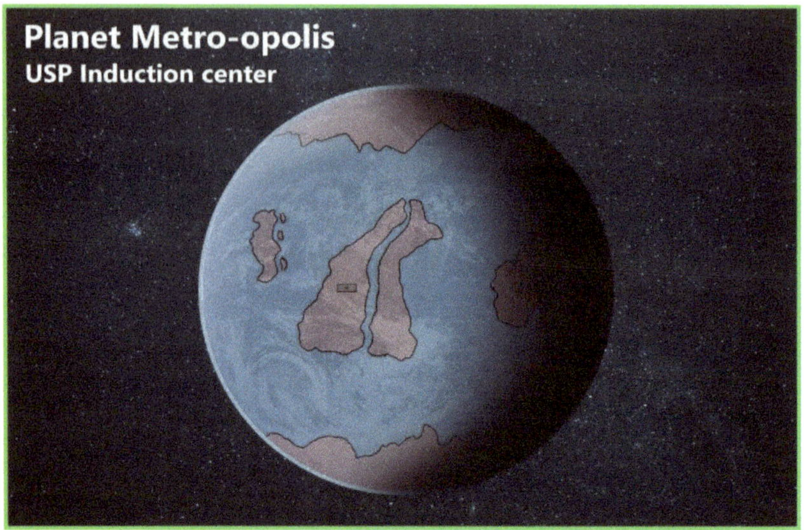

As they got closer to the planet, they followed Erik's directions towards the side of the planet where the Induction Center is located. Once they were within about 50 miles from the facility. Abran looked out the window and saw a huge, sprawling building that must be miles in length and said, "Damn that's huge! That's got to be it!" Then Erik heard a voice over the radio that was demanding! "Unidentified craft hold your present course; an escort craft will be joining you shortly." Like usual Abran felt very un-easy about the situation. Both Erik & Scott reassured Abran by telling him {Don't worry Abran, all we have to do is comply with all of their orders for now and everything should be alright.}

Just like the voice over the radio said, a small craft emerged from the facility and flew up next to them. The pilot's voice from the small craft came over the radio and said, "If you're here to visit the Induction Center then follow me. Otherwise, I'll guide you to a safe distance away from here."

Erik replied over the radio, "We're here for the Induction Center. We want to join the USP. So, we will follow you." So, they continued to follow the small craft. The pilot's voice from the small craft spoke over the radio saying, "Do you see that platform down there that says 43 on it? Because that's where you shall land. Once you've landed, a group of soldiers will be there to assist you. Other than that, I'll say welcome!"

With a delicate touch Abran precisely and gently landed the ship. Scott said, "Damn, Abran you've gotten so good at flying this thing. I couldn't even tell that we touched the ground." As soon as they landed Erik was gonna open the door, but Scott stopped him for a minute to say, "Hold on a minute guys! Well, we're here and if we've ever needed luck or a prayer, now is the time! Now, that we've reached the front gate of the lion's den. Things can either go one way or the other. All we can do is hope and pray that it goes our way." After that Scott signaled to Erik to continue opening the door. Then the door slowly opened, and just like they were told there was a small group of soldiers waiting for them. The brothers walked down the ramp to meet up with the waiting soldiers. Scott said, "Hello." The lead officer also said," Hello. So, I was told that you guys are here because you want to join the USP. Then in that case you three will need to follow me. There are several phases to undergo before you can call yourselves space police. The USP are some of the best officers in the known universe."

They all started walking towards the massive building that seemed to go on forever, the size was almost breathtaking.

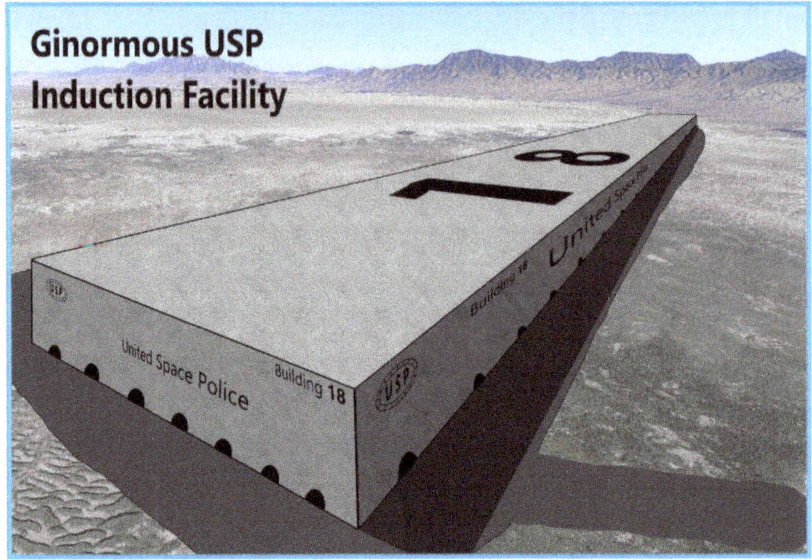

Once they entered inside the brightly lit, massive building through the front door. The brothers looked around and were astounded at how huge this building was. The building was 7 miles long, 2 miles wide, and about 400ft. tall. It had an open floor plan with a bunch of smaller buildings scattered around inside. Overhead the brothers could see what looked like thousands of robotic drones flying all about carrying small packages and other things. They also saw dozens of trolly cars suspended on hanging rail systems, carrying passengers between the tops of the taller buildings. On the ground there were dozens of soldiers all marching around in unison. In the distance they could see what looked like huge cranes lifting & moving space craft and other large vehicles around. The whole building was a buzz with activity. They followed the soldiers to a train station where there was a train waiting for them. The train looked just like a subway in a metro station.

They all stepped inside the train, and the doors closed. Then the train took off slowly and rapidly picked up speed. Within 20 seconds the train had reached its top speed, which was around 150MPH. The train had to go fast in order to get anywhere in this building in a timely fashion. Within about a minute they had already arrived at their stop. Only one of the soldiers got off the train with the brothers, he was going to assist them with showing them exactly what to do and where to start.

After all of the dozens of passengers exited the train. Then everyone who was waiting to enter the train was allowed in. Then the doors closed, and the train took off. As they were walking together the soldier explained the different phases that they needed to complete in order to become USP.

Everything starts at:

Phase 1: Admissions :

An initial identity check is performed, and you're admitted to start the Induction process

Phase 2: Medical Exams :

You're disinfected, inoculated, and have a thorough medical exam which will assist with Identity confirmation.

Phase 3: Background Check:

Your assumed Identity is ran through all the Intergalactic data bases for a final confirmation

Phase 4: Assignment :

Your job is selected, determine duration of training needed for your job, select living quarters, and given uniforms.

Phase I: Admissions

As soon as they walked into the admissions dept. a computerized voice told them to place their right thumb on a detector pad. The voice stated, "We must ascertain your assumed identity, before we can proceed." Scott was the first one to place his thumb on the detector. Once he did the voice said, "Your name is: Scott Rivera and your home planet is Tiplura located in solar system #1, in the Neptorian Galaxy, in super quadrant 368. Seeing that everything else checks out for now you're now ready to go on to the next department." Then Abran went next. The voice said, "Your name is Abran Almaraz, and your home planet is Grandlisto located in solar system #2 in the Reptiless Galaxy, in super quadrant 368, Seeing that everything else checks out you're now ready to go on to the next department." Then Erik was the last one, the voice said, "Your name is Erik Almaraz, and your home planet is Insitiania located in solar system #3 in the Golden Galaxy, in super quadrant 368, Seeing that everything else checks out you're now ready to go on to the next department."

Phase 2: Medical Exam

They only had to walk about 50 feet until they arrived at the medical station. Once they walked into this department, they were greeted by a technician. He stated, "Welcome, here we check your overall health and vital signs to see if you can handle the rigors and demands of your training. We also double check your identity using the data that we collect from your medical exam such as a DNA sample, retina scans, dental exam, and fingerprints to make sure that everything matches. Then you must be disinfected and inoculated. After which you will undergo a thorough medical exam, once you pass everything then you have officially been inducted and will start boot camp which lasts 9 weeks. Boot camp is not purely physical, it's mostly police training where you will most likely train for the job that you signed up for. Just because you sign up for a job doesn't mean that you'll automatically qualify for it. So, everyone has to be pre-qualified. Once you start boot camp, you'll have a lot of physical training along with classroom education. As time goes on, it'll be more time in the classroom and less physical training. Once you've finished boot camp and graduate, you will be fully trained and ready to start your job. Is all this understood?"

The Brothers all nodded their heads in agreeance to show that they understood.

The technician pointed them to the start of a 6 foot wide 400 foot long slow-moving conveyor belt. Then they were ordered to take off all their clothes and wait until the 2ft circles painted on the conveyor moved in front of them. Then one at a time, they each stepped onto one of the circles.

This spaced everyone precisely 12 feet apart so each process could be performed individually.

One by one they entered the disinfection area and were ordered to raise their arms, hold their breath, and tightly close their eyes for about 30 seconds. Then they were sprayed with a chemical from all sides and then rinsed with warm water. Then they went under a giant blow dryer to be dried off. Next, they went under a shrouded area and were told to keep their eyes tightly closed while they were irradiated with quick pulses of bright ultra-violet light from all sides for further disinfection. Then a medical technician walked onto the moving conveyor to give them medical gowns to wear until they were done with the entire process.

Next, another medical technician walked onto the conveyor to administer the inoculation shot. The shot was initially quite painful but is a necessary step in the process. Then lastly, they reached the end of the conveyor belt.

They were told to take a seat while a medical doctor begins the exam. He starts with taking a blood sample and checking their blood pressure. Then he runs a small scanning devices over their bodies checking for any abnormalities. Next, he has each one of them open their mouth to insert a test apparatus and bite down on it which instantly does a complete dental exam. Then he checked their eyesight using what looked like a small high-tech flashlight with multiple colors. Then each one of them stepped on a scale to check their weight. After everything was done, he tells them that everything looks good and hands his clipboard to the last technician and walked off.

Then finally the last technician finished entering their information into the computer. Then he tells them that he will be sending a transmission request to each of their home planet's data bases. Once he receives the correspondence, he will be able to compile all their records and medical data together using a computer algorithm. After which the computer will create a comprehensive identity for them that will be accepted throughout all the universal data bases. This identity will be considered undisputable and **final**. Once he sends the transmission the brothers won't need to wait at the medical department any longer, because it'll take many hours to receive the requested info. After which, an officer walked up to the brothers and handed each one of them a uniform. Then he tells them to change and points towards the changing rooms. Once they had changed, The Officer tells them, "That's better. Well now, I'll show you to your living quarters. Which are located in the recruit barracks, follow me."

As they were walking the officer told them, "It might take some time for the transmission of the data to get back & forth from your home planets, even when using the space highway to greatly speed up the transmission of data around the universe. Since your home planets are many millions of light years away, it'll probably still take around 11 hours for all the requested data to get back & forth. It might sound like a long time, but if the data was sent the old-fashioned way it would take **many** millions of years to get the answer, and no one could wait that long."

They walked into the recruit barracks, which was a large building across a small walkway. Once inside they went up a flight of stairs and walked down a long, dimly lit hallway passing many doors and they stopped in front of door #19. The officer waved a card over what looked like a keypad door handle which unlocked the door. The officer opened the door and said, "Well, here it is. This is where you're gonna spend the next 9 weeks of your lives, assuming your information checks out. When we get the confirmation data from your home planets, I'll come back to notify you personally. Until then make yourselves at home." Then he turned and walked away. Once he was gone the brothers closed the door. Then they started organizing their things and making their beds. Abran said, "Man, I sure hope when they get the communications back that the life hack turns out to be complete for everything." Erik said, "At this point Abran I really believe that everything worked out if you know what I mean. After we finish our 9 weeks of training we'll be space police." Inside their room was a computer terminal used for accessing the info-net. Erik went online to look up possible jobs available. After a few minutes of searching, they found a job called a long-range patrol officer.

After they read the description of duty it sounded ideal for them, allowing them to travel all over the quadrant. They all agreed that this job was the one that they wanted to train for.

After the 11 hours had elapsed the officer came back and rang the doorbell. The brothers knew who it was and were very nervous about the results. Scott went to open the door to reveal the officer standing there.

He said, "Greetings, I've come to tell you that after carefully investigating your identities and backgrounds, everything checks out and you're now officially USP cadets, welcome to your first day of training. Training will be conducted in 8 hour intervals meaning 8 hours on & 8 hours off. This will accelerate your training, but it'll still take 9 weeks. This schedule structure will compress a 12-week schedule into 9 weeks. I must warn you that the first 2 weeks will be pretty tiring, so try to get adequate rest which should be pretty easy because those beds are very comfortable. During your on time there is one 30 minute mealtime, so you don't have to use your off time doing that. During your off time you can rest and do all the personal hygiene that you need as well as recreational activities.

As of now you have 12 hours until you must report to physical training 102 at precisely 08:15. Just follow the other recruits to the closest train station. The train is programmed to take everyone directly to the training area. There will be over a 100 other recruits all going at the same time, so you can't miss it. Well, that's all the information that I need to give you, so good luck, and good day to you all." And with that the officer turned and walked away.

Abran closed the door and was the first one to celebrate by saying, "We did it everything worked, I can't believe it!!!" They shared Abran's joy, but Scott quietly started thinking about their future [So far everything had been successful til now. Even when we complete the police academy, then what? Can we go back to Earth? After all the rest of our family is still there, or maybe we'll just keep exploring. Yeah that's it we'll just explore everywhere the space highway will take us. Go to every quadrant and see if we can find the boundaries of the known universe.]

Training Weeks 1 & 2:

Both of these weeks were pretty grueling with lots of physical training. Each day was a complete challenge. One of the most tiring aspects about the first week was having to go through different rounds of qualifying trials. During each round, they would give it 110% of their best and hoped that it was good enough. At the end of each day, they would return to their room and go to bed totally exhausted. Towards the end of the first week, they were told that they had finally been qualified and got the job that they applied for. The following week their classroom curriculum along with their weapons & pilot training were tailored to match their job requirements.

Training Weeks 3 & 4:

During these two weeks the physical training was a little less exhausting and there was a lot more pilot training. They were disappointed to find that the USP ships were not nearly as good as their ship, but at least they were easy to pilot. Their piloting instructor noticed how knowledgeable they were about the ship's controls and praised them for this fact. During week 4 they enjoyed less physical training, even though it meant more time in the classroom. They were very grateful that there wasn't any homework which allowed them to utilize more of their off time for enjoyment and rest.

Training Weeks 5 & 6:

The physical training had kinda turned into just a daily warm up. After a bunch more pilot training they had been certified to pilot basic patrol craft and started to tag along with other patrol officers shadowing them for more training opportunities.

Training Week 7 :

The physical training was still just a daily warm up. After which a good portion was having to endure long hours in the classroom, learning local & universal laws and how to enforce them. Also, when they spoke to their patrol craft's computer for assistance, how to phrase the question to get the fastest and most exact answer. Followed by more pilot training.

Training Week 8:

The brothers were allowed to go on some short patrols and even make some routine traffic stops. They always had to travel with at least one experienced officer. They were never allowed to travel very far. They started to receive training to pilot larger transport ships. They learned that once they graduated, they could apply for very long-range patrols or maybe even a detective. Then they would be able to patrol the entire quadrant and beyond. Normally patrols are limited to the solar system closest to their base. Medium range patrols cover the galaxies for which the base is located in and focus on the solar systems that are the most inhabited.

Training Week 9:

The brothers wondered if going through all this training was worth it. It has been an exhausting training schedule. The brothers worked very hard and tried to learn as much as they could. With high ambitions to graduate they had been certified in patrol craft and transport ships. After seeing that the brothers had been so dedicated and loyal to their training. Their Sergeant, Mr. Mcdouglas recommended that the brothers should try to become detectives. If they succeeded, they would be allowed to travel all throughout the quadrant, and beyond which is a freedom that they wanted. The Sergeant also said that they could use their own ship if they wanted. It would have to be fitted with all the necessary hardware and markings in order to be part of the USP fleet.

Secretly the brothers still had plans to return to the Great Salt Asteroid to pick up Seth. Hopefully he'd still be there, but knowing that Seth probably got his money and bought his own ship. Then he'd almost certainly be heading straight towards super quadrant 92 to try to rescue his friend Jesús from the Prison Planet.

Graduation Day :

Throughout their training, the brothers kept to themselves, of course being friendly and respectful but remaining inconspicuous to help conceal their new identities. They were still a little paranoid about any slip up. At the start of this day, they were awakened by their drill Sergeant. As he walked by their room and rang the doorbell. He then inserted a small piece of paper into their mail slot. The paper was a program guide for the graduation ceremony.

After the brothers saw the program guide, they realized how unnecessary it was for them to attend the ceremony. They knew that no one was gonna come to see them graduate. We are only here to complete our {Life hack, *Said in a whisper*}. Abran said, "Instead of going to graduation we could use that time to start our job and maybe go out on a patrol. Remember we don't have to be space police forever, just long enough to seal in our new life." After this discussion they decided to go to the administration office to see if they could be excused from participating in the ceremony.

When they got to the administration office they asked to speak with the on-site commander. Erik explained to the commander why they didn't want to attend the graduation ceremony. The commander said, "Ok, I guess I will excuse you guys from the ceremony. Therefore, as of now you don't have to go. So anyways, here's your diplomas. Why don't you guys head down to dispatch. I'm sure they'll set you up with some sort of short distance patrol. That way you can start working right away." After the brothers heard that, they headed straight to dispatch. Once they got there, they spoke to the woman at the front window. She asked them if she could help them. Abran spoke up saying, "Yeah, we're looking to see if you could set us up with a patrol route. How about something local." The dispatcher said, "Ok, sounds good. First, I need for you guys to place your thumbs on the reader." After she checked their credentials she said, "The only local route I have is #412. Oh wait, I see that Mr. Parker is assigned to that route. Well, I don't think he'll mind if I give it to you guys, after all he usually complains about this route. So, I'm sure that he won't be upset to have another one, but I must warn you that the route you're getting is pretty busy. So, you're gonna have your hands full. Well, as of now this route is yours, so good luck!"

Chapter 14 "New Acquaintances, old friends remembered"

The Brothers all boarded their assigned USP ship to begin their patrol. After they took off and had been traveling for about an hour, they received a distress call from a ship about 10 million miles ahead. The ship indicated that it was having engine trouble, therefore it was adrift in deep space. The brothers set out to investigate and offer assistance to the disabled craft. After traveling for about 11 minutes, the brothers were now within 500,000 feet away and getting closer. They received a radio transmission from the disabled craft. The passengers were requesting to be towed to the nearest station where they could get the repairs they needed. Erik said over the radio, "We're United Space Police not a towing service, we'll place a call for you to be towed." A young girl's voice came over the radio saying, "Could you please! We can't wait for a tow because our life support system is starting to fail." Then Abran spoke up saying, "Hey Scott, their ship is too big for us to tow. We'll have to take them aboard in order to rescue them."

Then Erik spoke, "Hey guys, I just completed a scan of their ship and there's only two passengers on board. We should be able to accommodate them easily. Once they're safely on board, I'll call for a tow transport to retrieve their ship. The nearest station is pretty far away. So, I'll see if I can find a local towing service to help." Scott spoke through the radio telling the passengers, "Attention passengers of the unidentified craft. We will call for assistance with towing. We do not have the ability to tow you and cannot attempt any kind of repair right now. Your craft will be towed to the nearest USP outpost and there it can be repaired. Until then we will take you both on board our ship. Once we get to base you will have a normal identity and background check, so don't be alarmed this is standard procedure. Please allow us to board your craft." The passengers agreed to being boarded.

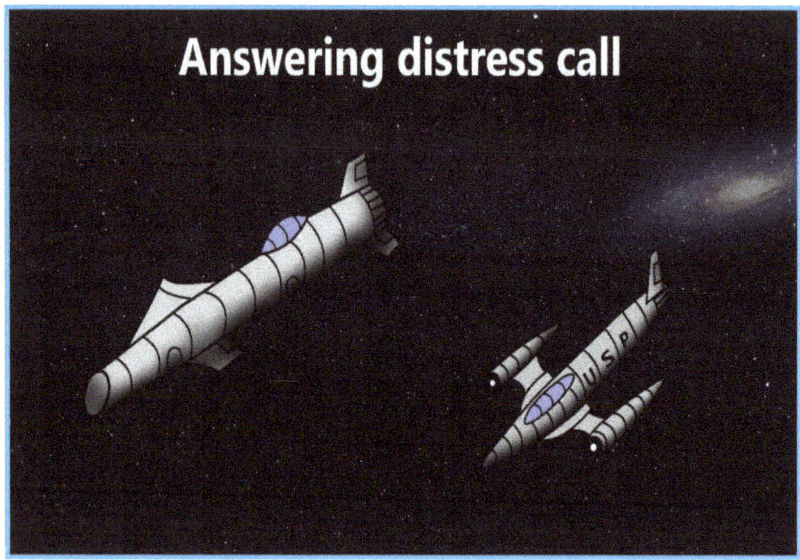

Once the brothers had steered their ship next to the other ship. They started the docking procedure by extending the airlock tube. As soon as the airlock tube had tightly sealed itself over the other ship's door hatch and was pressurized with air. Then Abran opened the door to the airlock and grabbed his gun, he made sure it was set to stun. Then he carefully entered into the airlock followed by Scott. He got to the door and was the first one to enter into the other craft. He was expecting to find the two crew members waiting by the door, but that wasn't the case. Instead, when Abran put his head into the craft and yelled out, "Hello! Anybody hear me, we're here to rescue you just like we spoke on the radio. You will not be harmed!" Then Abran heard someone's voice from down the hallway.

They said, "We're up on the bridge!" Both Abran & Scott cautiously walked down the hallway towards the bridge where they heard the voices coming from. As the brothers got to the bridge and peeked through the door threshold. They drew their weapons, just in case of a possible ambush. They saw a boy & a young woman standing in front of the control panel. They looked like they were feverishly at work pushing buttons. They turned around to look at the brothers then they said, "Oh, there you are! come on in."

Scott said, "We're here to rescue you! So, it's time to go, your life support is running out. Whatever you're doing, STOP! it's time to go right now!" The young woman said, "We're trying to get the life support working again, but nothing seems to help!"

Scott said, "We know! listen, we got to go NOW! Because the fact that we cannot repair that system right now. So, let's get going!" That's when the young woman & boy stopped what they were doing and followed Abran & Scott back to their ship. Once they all got back in the brother's ship Erik closed the door. Then he disconnected and retracted the airlock tube. The two ships started to drift apart. Abran engaged the engines to head back to base.

Scott started talking to the young woman & boy. He asked them their names and how old they were. The boy answered, "My name is Chris and I'm eighteen." Then the young woman answered, "My name is Nicole and I'm twenty four." Then Scott asked them where they are going? They both answered saying, "We're heading to Marcus 12." Scott said, "You know that's pretty far away from here. Anyways, what are you going there for? Because there's lots of other destinations that are closer and better." Nicole said, "Well, I have a really good friend that I grew up with, who lives there. She might still be there, or somebody there might know where I can find her." [Abran felt some strange intuition about this friend that Nicole mentioned.] So, he asked Nicole what her friend's name is. Nicole answered saying, "Her name is Kira… Kira Xena Jones." The brothers all couldn't believe their ears when she said that. Erik said, "Kira! We know Kira!" Nicole responded, "Really? Oh my god that's impossible! How do you know Kira?" Erik replied, "We first met her on the outpost base on the edge of the milky way galaxy. She traveled with us to the planet Nameka where we encountered her other friend, Bradley. They both traveled with us to the planet Aqualaysha. While we were there we encountered Seth Phenix, which is another friend who helped us a lot. As we were leaving Aqualaysha Seth requested to go with us. Both Bradley & Kira went in his ship together heading in a different direction. During our journey Seth requested to stop at the Great Salt Asteroid for some business he needed to do there. We had to leave him behind because of things we needed to do. We plan on going back there to pick him up. We also plan on helping him rescue his friend Jesús."

When Nicole heard this, she exclaimed, "Seth Phenix & Jesús Anaya! I can't believe it; we know them too! We met them both on the Great Salt Asteroid. They were there on some mission, but we never found out what that mission was. They helped us with our ship which was having some problems. Our ship's computer couldn't communicate with the engines therefore telemetry couldn't be established. Since Seth is so friendly, he started talking to us and found out that we needed help. After hearing our problem, he recommended his friend Jesús who is gifted at that kind of stuff. He invited Jesús to look at our ship, and within minutes Jesús had figured out what was wrong. The problem was kinda simple, he reset the computer, and changed the frequency to synchronize everything. After that the engines started up with no problems. He made a few other small adjustments which made our ship run better than ever. After that Seth & Jesús were our friends and we promised that we would repay their kindness. After they had repaired our ship, we all went to relax in the recreational area and sit at a table overlooking the Grandview of the deep space gallery. As we were sitting there enjoying each other's company, and just talking about things. From out of nowhere 4 USP officers came walking up to our table. One of the officers went right in front of Jesús and told him that he was under arrest for the charges of being a life hacker. They told him there were 2 counts of this charge. They told him to stand up and turn around. Then they handcuffed him, Jesús asked, "Where are you taking me?" The lead officer said, "We are taking you to the Prison Planet 4112, there you'll await your arraignment and trial." Then they walked him out in handcuffs. After that Seth asked us where we were going, we told him that we were heading towards Antilles in the Great Maritine Comet Belt. We showed him the map, and he asked us if we could take him to the planet Aqualaysha since it's along the way. He really wanted to go there, so we said OK. Then we all pretty much left right away. When we got to Aqualaysha he asked us to drop him off on one of the floating cities on the planet's surface, from there he would find transport elsewhere. After we dropped him off, then we all said our goodbyes and wished each other good luck. Then we got back into our ship, and headed off towards our destination, leaving him there."

After hearing Chris and Nicole's story, Scott realized that they were experiencing an unusual number of synchronicities, having a high number of mutual friends. The likelihood of these people who knew each other, either indirectly or through a friend was almost a disturbing level of coincidences. Scott was thinking [Everything that has taken place so far, has happened so perfectly. It almost feels like the hand of God is influencing all these events to make everything work out perfectly. Everything could be just pure coincidence or luck, but it seems too perfect for it to be anything else except divine intervention.]

Nicole & Chris both stated that, "After hearing that you guys know Kira, Bradley, Seth, and Jesús. We were originally on a mission to find Kira but now we've changed our minds. We've decided that if you're on a mission to help rescue both Seth & Jesús, we definitely want to be a part of it. After all they've done to help us, we want to help them too. Then afterwards if our rescue mission is successful you could probably help us locate Bradley & Kira using the resources of the USP. Therefore, we insist that you take us along to help rescue Seth & Jesús and we won't take no for an answer. We know that because you're USP that it might complicate things. However, there's got to be some way that we can tag along and help. I'm sure that you guys can figure something out."

Then Scott said, "You can probably tell by now that even though we are USP officers, we don't actually have any interest in performing our job. We are simply using our job title as a front to be able to travel around the universe. Our story is a long complicated one, which unfortunately we can't tell you everything right now. Therefore, after hearing your story I understand your request, and I've already kinda formulated a plan. So, this is what we're gonna do, we have to pretend to arrest you guys and take you back to base. We'll have to charge you two with smuggling, which will require us to take you to the Prison Planet. Don't worry everything is pretend, it's just so we can have a reason to go to the Prison Planet. Then along the way we'll stop to pick up Seth. Then all of us will travel directly to the Prison Planet to rescue Jesús. The other reason why we're going back to base is to get our ship. We only took this one because we had to for our first patrol, but this ship really sucks compared to ours."

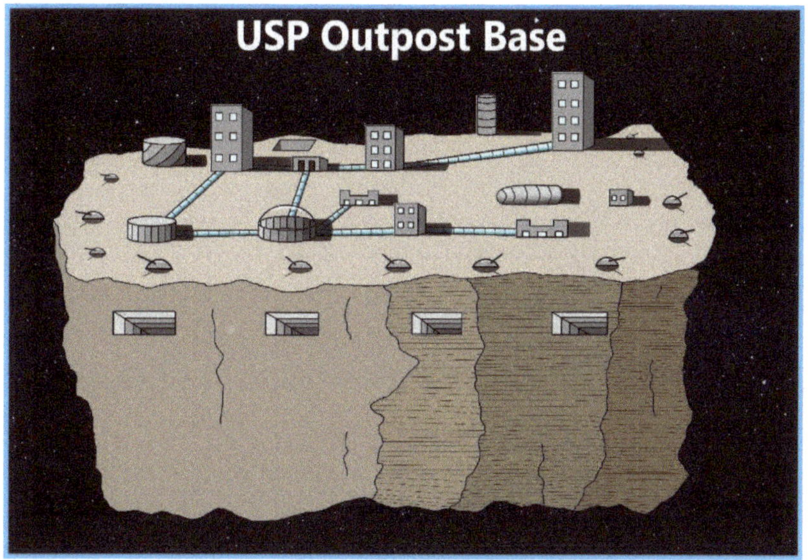

After they all agreed to Scott's plan they headed straight back to base and landed in the designated docking area. When they landed Erik deleted the ship's log to hide their encounter.

Luckily, the control tower personnel wasn't paying any attention to them. To help complete their scheme Scott told Chris and Nicole that," We're gonna have to walk you out in handcuffs. If you see any guards don't say anything to them. If they say anything to you, just say hello and not much else, don't have a conversation with them. After we're done here we'll be able to go pick up Seth if he hasn't already bought a ship. If he did, then he'll almost certainly be heading to the Prison Planet. Then maybe we'll be able to rendezvous with him somewhere along the way. However, the fact that he hasn't tried to communicate with us must mean that he's still on the Great Salt Asteroid waiting for us to pick him up. Therefore, our plan is based on that assumption."

Erik opened the door, and they walked down to the bottom of the ramp. A security guard just happened to be walking by and said, "Oh, I see you guys are back and you got a couple prisoners, good job! Where are you taking them?" Abran replied, "We're not sure yet, but they'll probably need to be taken to the Prison Planet for their arraignment and trial."

Then the guard said, "Alright, to the Prison Planet? Well, that's pretty far away. You know that's the base commander's decision. Anyways, be careful and bye." And with that said, he walked out of sight. As they were walking by the dock's central computer, which was just out of sight of any security cameras. Erik looked over and was stunned to see that it just happened to be logged in with the base commander sign on! Erik whispered to the others, "This is amazing, man this is gonna make things so… much easier using this incredible login. In fact, with this login I can create a Prisoner transfer for Jesús, that'll make rescuing him a lot easier. Unfortunately, that won't delay his trial and likely execution, but it'll make the rescue a lot more likely to be successful."

While using this high security logon Erik was able to create the prisoner transfer. He was also able to make clearance for Chris & Nicole to tag along as assistants instead of prisoners. After that Erik prescribed the journey that they would be taking which included the detour to the Great Salt Asteroid to pick up Seth and any stops along the way to get supplies. Therefore, this would not be considered an unlawful deviation of duty. Then he finalized all the vital information to make everything legitimate and indistinguishable from real orders. As soon as Erik finished entering these variables into the prisoner transfer log. The proposed trip was set, and the computer gave a reference #1128.

Scott was thinking [This just happens to be another one of those amazing coincidences of uncanny good luck. The logon we needed to create our mission just happens to be available. Once again, the hand of God seems to be guiding our journey.]

After the mission was set, they all headed towards their ship. Knowing exactly where to go, and also knowing that they had to adhere to a strict time limit. As they all got to the brother's ship, once again Erik opened the door, and they all ran inside. Abran & Scott ran right to the controls and started the engines. As soon as the ship lifted off, the docking bay shields turned off. Then Abran and Scott carefully guided the ship out of the dock. Once the ship was completely out of the docking bay, they sped off towards the nearest space highway on-ramp. Erik got the coordinates for their journey transferred to their ship's computer.

As soon as they had completed the data transfer, they were ready to go. They sped off as fast as they could heading towards their destination.

Chapter 15 " Friends Reunited"

As they left the USP outpost base heading straight for the nearest space highway on-ramp to get on route 447 that puts them on course to the Great Salt Asteroid. The only reason why they are going there is to pick up Seth. That's when Scott voiced out loud saying, "I sure hope that Seth is still there, I wish we knew if he was before we got there." After Chris heard that he said, "You know there is a way to find out, using the universal locator and it's usually pretty accurate. Any place that is closely connected to the space highway you'll have access to this ability. It usually doesn't take a long time." Then Erik said, "You need to show us this, because it sounds useful." That's when Chris said, "Good thing that you guys are USP, because this is only available to law enforcement. Anybody else who wants to use it has to hack into it." Then Abran said, "How come they didn't tell us anything about this at the academy? Maybe this is something that's only available to detectives?" That's when Chris went to work trying to enable this function for them. Luckily within about 5 minutes Chris had shown Erik how to enable and start using the locator. Chris explained, "Normally, all you need is the person's name and home planet and usually that's good enough to begin the search. If you have an ID# that's even better cause it'll greatly speed up your search. The computer will come up with all the possible matches, and you will have to select the match." Erik typed in {Seth Phenix}. That's when Chris said, "You still type in stuff? Why don't you just speak it?" Erik said, "I just feel like I have more control by typing stuff in." Then Erik finished typing the home planet {Schinické}.

As soon as he pressed enter, one by one a list of people and aliens with the same name & home planet along with a photo and basic profile appeared with their current approximate location. After about 5 minutes they saw a picture of Seth and saw that he was still on the Great Salt Asteroid.

Chris told Erik about the other features that the USP systems allowed. Which was to be able to contact him through the city's central computer. Therefore, they could communicate with Seth. Erik started the contact. Once they had connected to the city's central computer, he typed a message to Seth (Hey Seth we're coming to pick you up in approximately 5.5 hours. We are now USP, and we will be requesting docking bay 94 just like last time, so be there and be ready.)

They could only send a typed message because of the time delay. Any type of real time conversation with Seth just wouldn't be possible.

As the brothers, along with Chris and Nicole traveled along the space highway en-route to rendezvous with Seth. Their journey was uneventful as they arrived at the Great Salt Asteroid. Abran said, "Even though it wasn't a very long time ago. Man, it sure brings back good memories. Finding Seth should be easy since he should already know where to meet us, if he got our message." A voice from the traffic control came over the radio saying, "Welcome USP craft to the fun, fabulous & beautiful recreational heavenly body known as the Great Salt Asteroid. I offer complete assistance to law enforcement; how may I help you?" Erik answered back saying, "We're here to pick up a passenger, and we request to land in the underground docking area, specifically in docking bay 94."

The traffic control sounded intrigued saying, "Why do you request a platform so far away from the city? There are many vacancies that are much closer." Erik replied, "That's the docking bay that we were told to meet, so we need that one." Then traffic control said, "Very well, since that dock is currently empty, you may land there. Hope you enjoy your stay!"

As they got close enough to the underground docking area. The door opened up and they entered inside. They took notice that the same number of docking bays were occupied as before. Therefore, it felt a little silly requesting a docking platform all the way towards the end of the corridor. Especially when all the other ones in between were empty, but they were more concerned with finding Seth.

As they zoomed past all the vacant bays and got towards the end. They noticed that this time the lights were not dim and that haunted ,eerie feeling wasn't present this time. When they got in front of docking bay 94 the shields automatically went down, allowing them to enter and land. As soon as they landed the shields came back on, allowing the bay to pressurize with air. They were clear to open the door. Erik pushed the button to open the ramp. The ramp slowly opened, but before they walked out each one of them grabbed their weapon and walked down the ramp. They all walked over to the shuttle car door. Just when Erik was going to push the button to call for a shuttle car, they heard one arrive and stop.

Then Abran said with exuberance, "Alright! It must be Seth!" When the shuttle car door opened, sure as it could be, it was Seth! Everyone expressed great joy and said Thank goodness!

Seth looked over towards Chris & Nicole and immediately recognized them and said with surprise, "Nicole & Chris, oh my god I can't believe my eyes. I haven't seen you guys in forever! I'm so glad to see all of you, but I'm really happy to see you guys (Scott, Abran, & Erik) You came back to get me just like you said. At first, I didn't believe you, but now I'm a true believer and I owe you guys a big debt of gratitude!" Scott kinda butt in saying, "I hate to interrupt, but we really need to get going. We can catch up on old times along the way. I should tell you that we've all decided on a plan to rescue your friend Jesús and I'm sure that you'll want to be a part of it." After hearing that Seth said, "Oh heck yeah, of course I want to be a part of any plan to rescue Jesús. And I see that your plan worked to become USP, cause after all you're freakin' space police! Now that's something else man!" They all walked back up the ramp, in a brisk walk. Once they all got inside Abran & Scott went right to the controls and started the take-off procedure. Erik hit the button to close the door ramp and just as they lifted off the ground the docking bay shields turned off. They exited the bay and sped off towards the main entrance. Both Chris and Nicole watched all the empty bays whizzing by in a blur. They had to slow down a little to go through the underground docking entrance door.

As soon as they exited, they sped off towards the nearest space highway on-ramp. They were able to download all the space highway maps for all the super quadrants between here and there. Right away Erik began plotting all the routes to their next destination which was the Prison Planet located in the Shimmering Galaxy in super quadrant 92.

Erik calculated that they would have to cross through 15 super quadrants, traversing over 343,000,000 light years. So, even at the amazing speed of the space highway, it will still take about 17.2 days to get there. When Erik said that no one wanted to hear it. That's when Erik recalled an interesting fact that Kira had mentioned, which was, [Once you've entered the space highway, you can either turn your ship's engines off to save power, or you can use your engines to go even faster! There are physical limitations that stop you from being able to go light speed inside the space highway. Also, you must have very good shields to deflect any crafts from crashing into you.]

As they reached the on-ramp and merged smoothly onto route 4201 heading towards junction 3789. Erik continued to calculate how long it will take to get to their destination. When he looked up the sum of speed data and the limitations of the space highway. He marveled at the complete genius and flexibility of the space highway. One of the physical limitations of the space highway is if you go any faster than 100,000mps while inside the space highway the bonds between atoms start to break down. Both weak and strong nuclear forces holding the atom together are overwhelmed causing the complete disintegration at the subatomic level. Most crafts which travel on the space highway do not use any propulsion once they have entered. Once you're inside you are traveling around 5.8 quintillion miles per hour. So, there's usually not much need to go any faster.

Chapter 16 " Race to the rescue"

Now that they had embarked on their epic journey to rescue their friend Jesús, it was a race for time. As they all knew that once you're sent to the Prison Planet, you're pretty much guilty of whatever crime that you've been accused of. To go through the formal motions of being charged, and arraigned, then sent to trial, and finally sentenced, in that order. The chances of being found *not guilty* is almost non-existent, it is definitely a totalitarian government. The conclusion of guilt in this universe is almost definite; (*somebody has to take the fall.*) It kinda turns out if you just happen to be in the wrong place at the wrong time. Then you're gonna be the fall guy!

After researching about the possible speed gains while traveling through the space highway. Erik revealed his plan to accelerate them to 100,000mps for the longer stretches of the space highway (parts that are 2 million light years or longer) to shorten their trip as much as possible. By adding that extra 100,000mps to the space highway speed would make their total speed be an absolute mind-blowing 9,028,189,152,000,000,000 miles per hour. The brothers had almost become experts at navigating the space highway going from route to route and finding each junction, then merging effortlessly onto each one. Their planned journey consisted of going from the route that they're on now, and then to...

1. junction 3789
2. route 4202
3. junction 3790
4. route 4211
5. junction 9 the galactic junction to cross the inter-quadrant boundary line.
6. route 4214

There are a number of routes that they will be able to increase their speed to save time. Erik was looking up the exact procedure to safely increase their speed. Since the brother's ship uses Anti-Gravity-Nuclear-Plasma-Monopole for propulsion. Which gives them exquisite control and pinpoint accuracy within gravitational fields.

Since the space highway has both positive and negative gravity. Erik had to make sure that the engine's propulsion was the correct polarity. If they were to engage the engines with the wrong polarity, it would completely rip the ship apart disintegrating it and killing them all instantly! After doing some careful measurements they determined that they were on the positive side. So, with a high degree of certainty and much trepidation Abran slowly engaged the engines to about 50% of their maximum. Right away they felt the ship shutter a little. Erik thought [Uh-No, maybe I calculated it wrong?] That's when Abran announced, "Hey guys, according to the instruments we're already traveling almost 1000mps faster." That's when Scott said, "Whoa, make sure we put the shields on full power, right away! The space highway is dangerous enough, and that's without any additional speed. I'm sure that now after increasing our speed that'll make things, <u>a lot</u> more dangerous!" Right before Scott could finish saying that, Erik had already put the shields at full power. Even though they didn't feel anything there were indications that they were colliding with other crafts and just not feeling it. Their confidence rose along with their speed. Within about 5-6 minutes they had reached full speed. They were now traveling at 100,000mps. At this speed they could reduce their travel time by about 35%.

The 18.7 hour stretch that they were currently on would now be reduced to 12.1 hours. Their entire journey spanning 17.2 days should be reduced down to 10.5 days. Everyone was delighted to hear that their journey would be shortened so drastically, shaving six days off their trip is a huge savings. The sooner they arrive at the Prison Planet the better their chances of saving Jesús from being executed.

As they raced along the different routes of the space highway. Erik had already plotted their entire trip. He had downloaded all of the maps for the super quadrants that they'll be passing through on their way to the Prison Planet. After a short time of traveling everyone started to relax and get comfortable. The computer is capable of autopiloting the ship through most of the route changes, but human supervision is still necessary to make sure that the computer keeps them on course. Therefore, Abran & Erik had to remain on alert to assist the computer and double check the data. Scott, Chris & Nicole were discussing their journey at hand when Scott asked them saying, "Hey guys, do you mind if I ask you a little bit more about yourselves, and I'll tell you more about us."

Chris was the first one to speak saying, "Well, I'm originally from the planet Droid in the Golden Galaxy. I had a very good life with my loving parents until one time we were traveling between my home planet and the planet Insitiania. Suddenly, our ship suffered from a major engine malfunction and was about to explode! Our ship only had two escape pods, but the larger one got damaged leaving only the small one. My parents managed to place me into the small escape pod which was only big enough for me. They ejected it right before the ship exploded and only, I survived! Miraculously, I was found floating in deep space by a

USP patrol ship just like yours. Luckily, I survived inside this escape pod for two weeks without any food or water and with a limited air supply. Sadly, I don't remember that much about my parents, except that I love them and wish they were still alive. After that, I was sent to an orphanage on Marcus 12 and was assigned to live in the same home that Kira and Nicole were being raised in. I remember Kira was the oldest and very technically minded. Nicole was very empathic and friendly, which helped me heal from my emotional wounds. After some time, I began to immerse myself in the only activities that I felt were my best strengths, which are helping others and figuring out problems."

After Chris was done talking Scott asked Nicole the same question, right away Scott noticed that Nicole was fairly shy and soft spoken, but had a strong will.

Nicole said, "I love being with Kira, Bradley, and Chris because they are my family. Just like them I was orphaned as a baby; I was dropped off when I was just 5 months old at the local USP station on Marcus 12. The police conducted a DNA search for my parents and mysteriously they couldn't find any exact matches in this super quadrant. In fact, they could only find one perfect match: 175 super quadrants away. They felt that this was impossible that my parents could've got halfway across the known universe overnight. So, rather than pursue it they decided to place me into an orphanage until my parents could be tracked down, that is if they ever could be. So, that's some of my history."

Then she turned to Scott and asked him about their history and Scott proceeded to tell Nicole & Chris about their amazing adventures they have been through in the past few months. After hearing about what has happened to the brothers, Nicole & Chris were astonished!

Then Scott asked Seth about what happened to that code. Seth said, "After scanning it I couldn't figure out how to decode it. Without Jesús's help it was just too complicated for me to figure out. So I pretty much gave up for now. So, after that I was stuck waiting for you guys to come back and get me."

Every 24 hours they were able to travel through about 1½ super quadrants. Then right around the halfway mark Abran spoke declaring, "Hey guys there's something we almost forgot about…Supplies." That's when Erik sprang into action and started looking up all the possible destinations along the way where they could obtain supplies. He soon realized that the super quadrants they were passing through were sparsely inhabited with only a few worlds here and there. Some only had a single inhabited planet within a vast area, which was somewhat odd. Erik found the best choice happened to be in super quadrant 376 which had one inhabited planet where they could get supplies. Luckily, the space highway practically took them right to the front door. So, they would be arriving in just a few days. Until then their life support rejuvenation system can recycle all of their essential life support requirements for a while, but getting fresh supplies is always better.

As their journey continued, Abran kept thinking of the different scenarios that they might encounter when they got to the Prison Planet. Abran still doubts that the false prisoner transfer they created will be enough to safely reach the Prison Planet and rescue Jesús. He also figured that Jesús would probably be in a high security area, so rescuing him will be extra dangerous and difficult. Therefore, Abran was busy devising a clever plan in case this was true, so they would be ready.

As Scott was getting to know Chris and Nicole, he looked over and noticed how Abran seemed to be in deep thought. Scott has seen that devious, maniacal look before and knew what it meant. Scott knew that once Abran focused his mind on something he could figure out a plan that was fool proof. Secretly, Scott was a little envious of this natural talent that Abran had.

Suddenly, Scott realized that there might be a blueprint or a plan showing the Prison Planet available on the info-net, so he asked Erik to see if he could look it up. Erik went right to work searching all over but couldn't find anything. That's when Chris & Nicole overheard the discussion between Scott & Erik. Nicole said, "We overheard that you're looking for blueprints of the Prison Planet. The reason why you can't find anything is because it's listed by the planet name which is **Arcadia**. This information is kept very classified for obvious reasons of high security. Each inquiry is monitored, and the fact that you guys are USP, will probably make this inquiry be disregarded."

After Nicole said that Erik sent an inquiry, and shortly after that the download came available. Once it did, Erik downloaded it. As soon as the download was finished, he opened it.

The blueprints were very detailed, and they were surprised to see that it wasn't just one giant prison covering the entire planet. There were numerous prison compounds that are spread miles apart covering the entire globe. Each one incorporated all the systems needed to sustain every aspect of life and community.

Such as the cell blocks, medical facilities, food cultivation, judicial division, execution, coroner/mortuary, transportation, waste control, communication, maintenance, firefighting, and chemical control. Pretty much everything needed to sustain an independent society, except that this society's only industry is a military base that maintains and governs the prison compounds. After all of them reviewed and analyzed the blueprints, they agreed that Jesús was most likely being held in the high security execution ward.

Each execution is televised for several reasons, one reason is for witness. The other reason is for an example to show other possible criminals of what fate awaits them when they're caught committing one of these capitol offenses. There is a whole industry devoted to the execution portion of criminal punishment. The whole thing is designed to discourage any copycat criminals.

Erik typed in "Jesús Anaya" the computer searched for a few seconds, then low and behold a match. The only oddity they found is that he was being held in a fairly low security cell block. Which must mean that he either hasn't been to trial or he is considered low risk. There wasn't any more information available detailing Jesús's incarceration. Abran said, "Let's hope that he's still waiting for a trial and not simply waiting for execution. Then hope he doesn't get moved to a more secure location because we only have a few days and hours until we get there."

Chapter 17 " A brief stop in super quadrant 376"

After a little more than 5 days of traveling they finally entered into super quadrant 376. Which is where the one civilized planet that they should be able to purchase supplies. As soon as they entered into the super quadrant they immediately started slowing down, so when they got to the exit ramp, they would be able to exit. Abran disengaged the engines to start the slowing process. They were slowing at a rate of 1600 miles per second slower. At this rate it will take about an hour to get completely back to a neutral speed. After the hour had elapsed Abran announced that they had finally reached neutral speed and the exit ramp for the Dim galaxy was quickly approaching. Abran & Erik prepared for the exit that was approaching within 5 minutes. Like most of their journey, there was no room for error. As they got closer to the exit ramp Abran & Erik had to be on high alert to make sure they'll be ready to exit. Even though the space highway normally slows down when you enter into galactic space due to the change in gravity density. They expect an additional slowdown will occur right before the exit ramp allowing more time to safely exit.

Within no time at all they arrived at the exit ramp. On the 10,9,8,7,6,5,4,3,2,1 then they smoothly and seamlessly exited the ramp. Once they were out, they headed straight for the Dim galaxy solar system. Luckily, they weren't very far away from the solar system and planet Eazy which has the supply outpost on it. Scott said in a humorous tone, "Well, going by the name of this planet, we can only hope that this mission will be easy." Erik added, "And quick too."

Within 45 minutes they had reached the planet Eazy. Abran said, "Well, everything looks normal on the surface and according to the sensors, like usual it has breathable air, normal gravity, and atmospheric pressure. So, we should be ok." [Abran remembered one thing that Kira said that "All the planets that the space highway goes directly to are inhabited with some sort of civilization, otherwise the space highway wouldn't go there."]

To mention it, there is a whole history and timeline about the space highway and when it was constructed and who created it, but that's another story.

The closer they got to the planet's surface they noticed how green it was. The vegetation was very lush and dense with a lot of colossal sized trees.

Once they entered the upper atmosphere, their radio crackled to life with the sound of an automated female voice stating, "Welcome travelers! to planet Eazy, a tropical jungle paradise. We noticed that you're law enforcement and we don't get very many visits of your type. So, in that case if you're here on official business, you can check in at our law enforcement headquarters when you land. It's located at 25°longitude, 10°latitude. Otherwise, you can find all the information you need on our traveler's portal. That's all for now, enjoy your stay, see ya!" Abran said, "We need to make this visit very short and sweet." Erik located the supply emporium which was about 2000 miles away. After hearing that they took off heading straight for it. Everybody had to squint while looking out the windows at the lush, green, tropical, jungle paradise. It was a thick jungle, lit up by very, very bright sunlight. Erik and Chris were looking up information about this planet. One fact was that the star in this solar system is a rare type, it's a super bright blue giant. Which means that it produces about 21% more light than our sun but has the same amount of heat. Plant life just simply thrives in these conditions and that explains why the plants and trees are so big and healthy. Therefore, plant life completely encompasses the entire globe. One of the first things they need to get is sunglasses. Even though they only plan on staying here for a short time. Any duration longer than an hour can result in eye damage or permanent blindness, so proper eye protection is necessary. As they sped along and stared out the window, they were mesmerized by the profound beauty that this planet possessed.

As Chris looked out the window, he found himself daydreaming about his parents and wishing he could see them again. The few memories that he had of his parents were very fond. He wishes that he could turn back time, to a happy time because right now he felt a lot of sadness even though he has close friends, almost like family(Nicole & Kira).

He felt this hidden void within his heart, feeling a deep despair, an emptiness that felt painful. He found himself not caring if things got better and wanting just to DIE! He kept these feelings quiet, not wanting anyone to know or help, and not wanting it to pass. He felt some strange comfort in feeling sad.

Chris's daydream was broke by the sound of Scott's voice saying, "Well, we're here, is everyone ready? If not, then get ready!" As soon as Scott finished saying that, Abran announced that they had reached their pre-designated coordinates and it's time to land and get out. Chris was still looking out the window and saw that they were coming up to an absolutely gigantic tree. As they entered into the canopy and looked through the leaves they saw what looked like a giant tree house, or more like a tree mansion. It was quite large but was dwarfed compared to the tree. Scott was thinking out loud and said, "This place is definitely not what I expected it to look like. Don't get me wrong it's beautiful, but let's hope they have everything we need and let's make it quick!"

Erik had just finished analyzing the outdoor conditions to verify that it was perfectly safe, then he gave a thumbs up.

As soon as they stopped and landed Scott pushed the button to open the door. All three brothers grabbed their weapons and walked down the ramp followed by Chris, Seth & Nicole.

Even though they were shaded under the canopy of the giant tree the light was still kind of bright. They walked around the area surrounding the supply emporium observing the view and noticed how high up they were, well over 2000ft.

As they were looking down at the ground Nicole saw something big moving in the distance. The more she looked at it trying to figure out what it was, the more puzzled she became.

Finally, she determined that it was some really large beast roaming around on the ground. When she pointed it out to Abran, he said, "It kinda looks like a dinosaur or something like that." As everyone looked around, they noticed that there were large creatures all around. This whole planet was experiencing a prehistoric era such as the Mesozoic or Jurassic time of the dinosaurs.

Chris said, "Well at least most of the big ones can't reach us because we're too high up on this giant tree, but the smaller ones that can fly or climb can get us." After this realization they all briskly walked towards the entrance of the supply emporium. Abran got to the doors first and discovered that they weren't automatic, so he held them open so everyone could walk right in.

They all walked in and as soon as the door was closed it was dark compared to the outside. It took a minute for their eyes to adjust. After they could see again Scott was the first one to say, "Man, I don't know about this place it doesn't look like it's right for us." They looked around and noticed that this little shop was a lot bigger on the inside than it looked from the outside, because part of it was inside the tree trunk. Abran said, "I'll go look for a cart. Don't worry, I won't go too far." So he split up from the others.

As they walked down the twisted aisles which made this shop look like a maze full of plants and artistic sculptures. Scott was thinking [I guess the only people who would want to buy a plant here is a tourist or some interplanetary traveler. Otherwise, I can't imagine anyone who lives on this planet wanting or needing to buy a plant, after all this whole freakin' planet is a giant jungle.] As they walked down the aisles, they passed intricate bonsai trees and other nik nak type items. They continued walking and noticed that the items for sale began to change until they saw the items that they actually came here for. The presentation of the items was different than anything they had ever seen before, but somehow oddly satisfying. Then Abran came walking up pushing a little shopping cart to put their stuff in. As they walked by the items they needed they put them in the cart. However, they didn't see any sunglasses anywhere. So, Scott said, "Maybe I'll ask them, when we get to the counter to pay." Once they had all the stuff they needed they headed straight for the counter in the back of the store.

When they got to the counter Chris rang the little service bell and they waited. They didn't know what to expect, would it be a human, alien, or some other strange creature?

After about 20 seconds of patiently waiting Chris was about to ring the bell again but before he did, they all heard the sound of feet shuffling very lightly. Then from behind the counter a small curtain was pulled open and slowly out walked a short old man with a mostly bald head and a long white beard, wearing small glasses with round lenses. He was dressed in beige slacks and an off-white long sleeve shirt with the sleeves rolled up and most notably a small wart on the tip of his wrinkly nose. When he got to the counter, he spoke in an English accent with a raspy old man voice which definitely suited his looks. When he spoke, he said, "YES… can I help you?" Chris and Abran started putting the items from the cart onto the counter. Scott said to the old man," We would like to buy all this stuff, and some sunglasses please." The old man said, "Sunglasses, I stock them right here." Then he reached under the counter to grab a small box with sunglasses in it. He set the box down on the counter and said, "You'll definitely need these when you walk outside, otherwise you'll be blind without them." Everybody grabbed a pair of sunglasses; it was an easy choice because they were all identical. After that, the old man began to pick up each item on the counter and carefully read a small sticker on it. He read each one out loud, and typed it into a register. Then once he was done, he totaled it up saying, "A total of 10 Intergalactic units please."

Then he glanced up and said, "Oh, I see that you're USP, therefore that means you get a 10% discount. In that case it'll be 9 units please." After paying the old man, everyone helped load the stuff back into the cart. The old man said, "Thank you for your business and good day to you all." And he slowly walked back through the curtain that he came from and disappeared.

Everyone turned and quickly walked towards the front doors. When they got there, they quickly stopped as they looked through the front windows. Outside they saw what looked like a large group of mini T-rex dinosaurs, they stood about twice the size of a chicken. They estimated there were about 40 of them standing in the way between them and their ship. Abran commented saying, "These creatures look pretty mean!" Everyone seemed to agree with this assessment. The next question Abran asked was, "I bet if we ran out and started shooting, they'll probably just get scared and run away, what do you think?"

Then from out of the nowhere they heard a woman's voice that said, "That would be un-wise." Then a short, middle aged woman walked out from the next aisle. Scott was the first one to say, "Who are you?" The woman responded saying, "I'm sorry for listening in on your conversation, but I couldn't let you make a fatal mistake. Oh, by the way my name is Nora. And those creatures are called piranha dons, and a direct attack doesn't scare them away, instead it provokes them to counter attack. They smell you and they're hungry, so they won't leave. You can try a decoy because they're not very smart, so that might be one way of outsmarting them."

The first one to respond to this suggestion was Chris saying, "Why don't we try using a laser pointer, maybe they'll chase it just like when you're playing with a cat."

Scott replied, "That's a good idea but I think it's too bright outside for that to work." After hearing that Abran was able to turn the laser on his gun up to maximum brightness so it can be seen even in the bright light. He pointed it out the window a little past where the piranha dons were standing. The piranha dons saw the laser and started chasing after it. Then Abran pointed the laser on one of the piranha dons in hopes that they would attack where the laser was shining. Abrans plan worked, the piranha dons started attacking and had pretty much killed & devoured the unlucky one that he targeted! After that they smelt the blood on each other and started attacking each other! There was blood and body parts flying everywhere! During the frenzy, the brothers saw their best chance to escape. They put their sunglasses on and each one of them grabbed a handful of the supplies out of the cart and ran as fast as they could!

Abran ran out first, followed by Scott, Erik, Seth, & Nicole. Chris accidently dropped his sunglasses which delayed him from running. They ran out of the door carrying their stuff, and were unable to use their weapons with their hands full. Abran got to the ship first and managed to push the buttons to open the door, so it would be fully open to allow each of them to run inside. Since Chris was the last one to run out of the shop. The piranha dons turned their attention towards him and started chasing him. Nicole looked back towards Chris and shouted to the others, "Oh my god they're chasing Chris I think they're gonna get him!" Chris was the last one running, he turned his head to look back at the piranha dons to see them chasing him, which terrified him! With his adrenaline pumping which gave him almost super human speed to run faster! Chris got back to the ship with just enough time to run up the ramp! Erik closed the door right in the nick of time!

The piranha dons could run very fast, so by the time the door was shut a few of them had already reached the ship. Our friends could hear the piranha dons running all over the outside of the ship. Everyone took off their sunglasses and set the supplies down. Abran & Scott ran straight to the controls and started the engines, they slowly started to lift off. The moment they got about 200 feet away from the great tree they sped away at an incredible speed! The few piranha dons that had stowed away on the outside of the ship were hurled off and fell thousands of feet to their doom.

Within less than a minute they had left the planet Eazy and were heading straight back to the space highway on-ramp. Once they entered the space highway they sped off towards their final destination.

Even though they were on a galactic route they were still able to gradually gain speed. Within about 8 minutes they had regained over 80% of the speed before they stopped. Even though they were traveling at a fantastic speed, it will still take about 7½ days to get to the Prison Planet. Scott was delighted that they were able to obtain enough fresh supplies for the duration. In that case they wouldn't have to rely on the recycler as much. Since they were not planning on stopping or even slowing down until they reached their destination.

Once they got back on the intergalactic routes, they were able to reach full speed once again. During each 24-hour period they would spend vast amounts of time plotting and planning their methods. Scrutinizing all the details that they could see, to enter, rescue, and escape the Prison Planet successfully even if they encounter high security. The fact that they're USP who are usually allowed in areas without as many security protocols, makes success more likely without every move being questioned.

Abran quietly piloted the ship alone, by choice. Abran preferred to do this alone. He found that sharing this activity to be somewhat distracting. Only when an operation that required more than one person, then he'd ask for help. Otherwise, being alone and doing things alone was always preferred.

Everyone else was reviewing the Prison Planet blueprints looking for any security gaps and ways to maximize those vulnerabilities. To ensure that their prisoner transfer and rescue of Jesús will be successful.

Chapter 18 " The Prison Planet in sight"

The final day of the longest part of their journey had finally arrived. As soon as they entered into super quadrant 92 Erik advised Scott that they should start slowing down to ensure that they'll be at neutral speed when they reach the junction. Scott broke away from the group to go help Abran co-pilot the ship. They were currently on route 3330 heading for junction 3001 which will take them towards the shimmering galaxy. After studying the maps Erik realized that they were gonna have to travel through 3 galaxy clusters and 4 galaxies to reach the Prison Planet. There isn't a short cut, and this routing is intentional to reduce the chances of any escape. There is two checkpoints along the way to thoroughly screen everyone entering the Prison Planet.

Spread throughout the known universe are some 20-30 Prison Planets. Criminals, especially ones that have committed the most serious crimes such as life hacking, are sent to one of the Prison Planets.

The proximity alert sounded letting them know that they had 5 minutes until they'll reach the junction. They were expecting the normal slow down except Abran noticed that they were slowing down a lot more than usual. That's when Chris & Nicole both said, "There must be a checkpoint coming up before we get on the junction. The average visitor must come to a complete stop so they can search your ship to ensure that any persons or objects not allowed beyond this point, are stopped!

The officials at the checkpoints are allowed to use deadly force if they **want to**, but since you're USP they might just wave you through with only slowing to verify your credentials." After 4 minutes had elapsed, they were now only traveling a few thousand miles per hour. After coming from such an absolute fantastic speed, it now felt like they were almost standing still or even moving backwards. After a few more minutes they had slowed down to just a few hundred miles per hour which might mean that they'll have to stop at this checkpoint. When they saw the checkpoint it looked like a very well-lit border crossing. Chris said, "Well there's the checkpoint, let's hope we don't have to stop!" and Scott agreed.

Within less than a minute they arrived at the checkpoint. There were dozens of lanes to help process all the ships coming through the checkpoint. The brothers were directed to the lane designated for law enforcement which had a lot less traffic and was moving much faster than the other ones. They received a communication from the officials which said, "USP craft #1324287 please transmit your authorization code." Erik proceeded to transmit the requested code; saying, "I hope this works?" Within one very, very long minute the voice said, "Welcome, USP craft your prisoner transfer #1128 is expected, please proceed directly to the next checkpoint."

They went through the first checkpoint and junction. Then started to dramatically accelerate, within minutes they were back up to full speed heading to the next galaxy cluster and checkpoint. While on these smaller segments between checkpoints they wouldn't be able to accelerate beyond the speed of the space highway, because they wouldn't have enough time to slow down.

There was only about 4.7 hours until they'll pass through the Globular Galaxy. In no time at all the 4.7 hours came and went almost in the blink of an eye. They went through the Globular Galaxy without stopping and with minimal slowing and sped off towards the Twisted Galaxy. This high-speed space travel has become so mundane, that the hours seem to go by without much notice. Even though the fact of how dangerous every moment of traveling on the space highway actually is, that fact has been disregarded. The last checkpoint lies within the Spider Galaxy, there will be no stopping or slowing until then. The brothers were constantly aware that the time available to rescue Jesús before his execution was getting dangerously short. Erik ran another check on Jesús to see if he was still alive, and if so then what was the status of his incarceration.

They found that he was still alive and had already been through trial and was found guilty. He also has been sentenced for execution; therefore, he is scheduled to be executed 16 hours from now. After hearing this update everyone was worried about the time constraints being so tight, and once again they found themselves in a suspenseful situation. One of the only good things about the execution ward is that the security is relatively low and mostly automated. Since it is automated, these systems can be predictable, and that fact alone makes them somewhat easier to fool.

Once they arrived at the edge of the Spider Galaxy, they were also approaching the last checkpoint before they'll reach the Prison Planet. Chris and Erik were busily searching for any information about the checkpoint. After searching and searching they realized that the last checkpoint was so highly secretive that there wasn't any information available.

All Scott could say was, "Since we don't know what this checkpoint is… checking for. Let's make sure that all of our digital signatures are accurate and lined up, so we can avoid any time-consuming stops that could possibly ruin our carefully planned rescue."

They began to experience another involuntary slow down as they got closer to the last checkpoint. After their experience at the previous checkpoint, they kinda knew a little about what to expect.

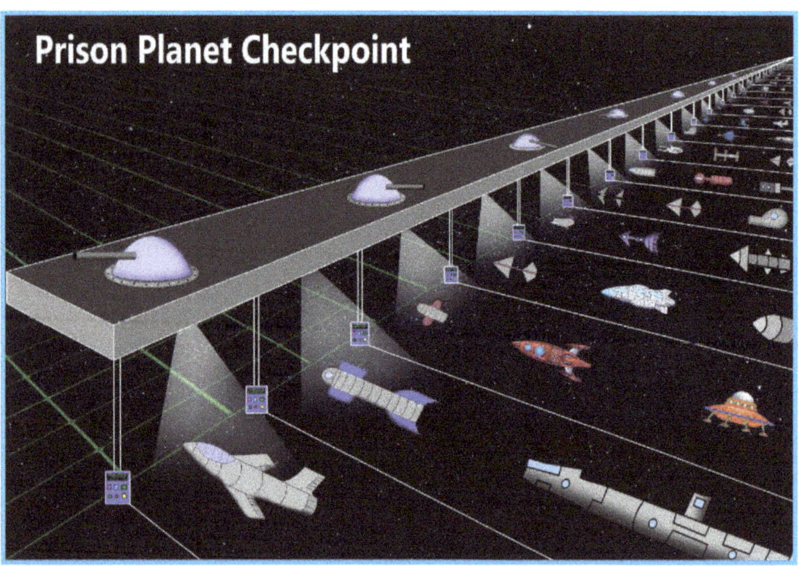

When they got close enough, they saw thousands of space craft stopped in front of what looked like a large toll bridge plaza or a border crossing. The checkpoint had at least a hundred bays to process a lot of ships at the same time. Most of them consisted of civilian and service crafts, only a small portion were USP. Everything was monitored by the highly armed security outpost. Just like before a voice came over the radio telling them to enter the lane for USP crafts. After they had come to a complete stop and began waiting for their turn.

Which usually takes between 3-5 minutes for each craft to be processed through the checkpoint. As they got closer to the front of the line there were only 4 crafts in front of them.

Then they saw one craft in the civilian lane next to them being escorted off to the side next to the security outpost. Out of curiosity they watched the craft being escorted, at first everything seemed ok. Then suddenly the craft started shooting at the security outpost as it sped away trying to escape! Quickly the large laser cannon from the security outpost fired a giant laser blast at the fleeing craft! The laser hit the craft in the center which caused it to break into large pieces! The pieces exploded with a blinding flash of light! Everyone stopped to look at the excitement of the craft exploding for a few seconds! Right after that everything resumed to normal as if nothing happened.

Then a small fleet of little drone robots exited the security outpost to clean up all the debris that could cause damage to all the crafts coming and going. Therefore, it must be cleaned up right away, plus it looks bad. After seeing the events unfold before their very eyes, they were shocked at the level of reaction and use of force that was applied. Not knowing what violation that the now destroyed craft had committed, but the level of force that was used seemed like it might have been a little excessive.

After 10 minutes of waiting. It was now their turn; they came up to the front of the line. A voice sounded through the radio demanding to board their craft, so they agreed.

Then out from the security outpost came a boarding chute which extended and connected itself to the outside of their ship, allowing the guards to come aboard. Abran opened the door and standing there were 5 guards waiting to come on board. Once the guards entered their ship, the lead officer stated, "Since there is no record of this craft ever accessing the Prison Planet 4112. Even though you're USP with a verified prisoner transfer #1128. This is a standard procedure to verify all personnel aboard this ship, before you can be released. So, we'll need a fingerprint, retina scan, and D.N.A. sample from saliva. Once you guys and your ship have been verified and confirmed to be authentic. After everything is verified then there will be no more inquiries. So, when you leave, you'll just go through each checkpoint without having to stop." They allowed the officers to collect the samples needed from the group. Within about a minute or so everyone's sample came back as verified and approved, except for Chris's. His sample took longer than everyone else's.

Three minutes turned into ten, then finally the officers asked Chris to come with them for a more in depth identification process. After Chris went with the officers Scott started thinking [This is bad! Really bad! Not only for Chris, but for all of us and our mission too. I can only hope that it's not what it looks like!]

After another very long and awkward 10 minutes, finally Chris and the officers came walking back, they were all laughing about something. When Scott saw this, he knew that everything must be ok, but he wondered what the inquiry was all about.

The lead officer walked up to tell Scott, "Hey man we're really sorry for the delay, but Chris's identity was really difficult to verify because there's 3 other individuals that are really close matches, but everything checks out fine. So, you guys are all done and free to go as soon as you're ready." Then the officers said goodbye and they all walked back into the docking chute. The docking chute detached and pulled away from their ship, and they were now free to leave. The brothers all said, "Man that was way too close Chris, we all thought you were toast!" Scott said, "Well guys I think that's the last hurtle, let's move!" And with that, Abran re-engaged the engines, and they sped off. Erik gave a time read out saying, "We are now only about 2.7 hours away from the Prison Planet" After hearing that Scott calculated that Jesús had about 11.5 hours left to live and counting down.

During this travel time everyone went over all the known facts of the impending rescue. They rehearsed their plans and discussed the known & unknown facts. Along with the worst-case scenarios that they are more likely to encounter. They decided that it would be better to split up into 2 teams.

Team 1: Scott, Abran, and Erik

Team 2: Chris, Seth, and Nicole

Each team will have a good mixture of talent, technical expertise, cunning, and intuition. Team 1 needed to be all USP officers so there won't be any questions about their credentials while picking up Jesús. Just that fact alone is what qualified each team member. Then lastly, each team member will be responsible for themselves. If any part of their plan should go wrong and any team member were discovered or captured.

The rest of the team(s) may not be able to spend precious time rescuing anybody and doing so would add risk for everyone. In that case the team or individual who gets caught might have to be left behind so that the rest of the mission could still be successful.

Leaving someone behind is the absolute last resort and they would do just about anything to avoid this. Everyone felt the tension when they thought about all the ways things could go wrong.

Then Nicole spoke up,[Nicole has been very quiet up until now] She said, "Hey guys lighten up, remember that everything has went very smoothly til now. Also you guys are United Space Police and our story all checks out. So, I think everything will go fine. So, relax a little because we're gonna need clear minds. Maybe take a short nap. In other words, just relax and chill out!"

After Nicole's pep talk everyone seemed to be a lot more at ease. There was a calming effect that came over everyone during the remaining time.

Suddenly, everyone jolted when they heard the proximity alert sound. The alert signified that there was 2 minutes until they'll reach their destination. Now everyone was clearly nervous, and for good reason. They had reached the off ramp and the countdown began until exiting 10,9,8,7,6,5,4,3,2,1 then they exited the ramp...

Chapter 19 " The Prison Planet rescue"

Poised right before them within a mere 250,000 miles away was the diabolical Prison Planet. A small planetoid noticeably smaller than Earth and was lit up by a nearby star. As they got closer to within a 1,000 miles away, they could see a lot more details. When they looked out the window, they noticed that it appeared to be somewhat rust colored with a thin smoggy atmosphere. There was an almost non-stop flow of ships coming and going from the planet's surface through one of dozens of entrance hubs in the force shield that surrounded the entire planet.

Right away they noticed that this planet is highly militarized. Each prison compound is basically a military fortress, with a prison yard next to it. The Prison Planet can best be described as one giant military base that manages the multitude of prison compounds covering the entire globe.

Each one has different degrees of security and prisoners who have committed similar crimes.

The night side was by far the most sinister looking. It had what looked like giant bonfires spread miles apart randomly covering the entire hemisphere. The bonfires glowed ominously with a devilish looking red hue. On the other hand, the day side was in its own way very un-welcoming. They could see the disturbing sight of what looked like the outlines of giant prison yards with the movement of thousands of prisoners pinned up inside. As they got closer the radio came to life announcing, "Welcome USP craft, a landing pad has been selected for you. Your prisoner transfer #1128 will commence when you arrive at cell block G12F30012E." Once the brothers heard this, it put them at ease. The control tower transmitted all the details about their prisoner transfer and their designated landing platform. After viewing the transmission, they learned that their landing platform was #327. When looking at the map they noticed that the landing platform was about 50 miles away from the execution ward. They wished that it was a little closer due to the fact that time was running low. Abran guided the ship above the landing platform and while trying to land he found it difficult to control the ship. He made the comment saying," This is weird, the ship is acting like the gravity is fluctuating. It's actually bucking a little bit." Immediately Erik looked at the gauges and saw that Abran was right, he determined that there's a tractor beam that's messing up the engine controls. That's when Chris said, "Since there's a tractor beam you need to turn the engines off, trust me!" Reluctantly, Abran complied and turned the engines off.

Then just as Chris said the tractor beam took control over the ship and landed it safely & gently.

Once they were ready to exit the ship, this time Abran was the first one to get to the door. He reached to grab his gun but stopped himself. Nicole saw that Abran was having difficulty deciding whether or not to carry a weapon. Therefore, she asked him, "What's wrong Abran?" Abran said, "I wanna bring my gun, but I'm not sure if it's ok to do that here." Nicole stated, "Since you guys are USP, you definitely need to take your weapons. In fact, it would look weird if you didn't."

Scott chimed in saying, "We'll also need to take a set of personal communicators. So, we'll have a private and encrypted way to communicate with each other when we split up. As you already know that each team will be responsible for different tasks to ensure a successful mission. Team 2 will be responsible for making sure that the tractor beam is disabled. Then if needed you'll have to hack into the computer system to restructure the return mission. So, leaving here will be possible. Me, Abran, and Erik will go to the prison compound where Jesús is being held to complete the prisoner transfer. Then if the mission is successful, we'll meet back here, at the ship."

As soon as they all left the ship, team 1 started walking towards the nearest train station that would take them to the prison compound.

Team 2 started walking in the opposite direction towards the tractor beam controls and administration building which was only about a half mile away.

As Team 1 arrived at the train station and waited for a train to show up. Luckily, at that moment a train came speeding into the station and quickly came to a stop. The doors opened and some passengers came rushing out.

Then they walked inside the train and took a seat. As soon as everyone was seated the doors closed and the train went speeding off. Within no time at all the train had reached nearly a 100mph. Erik saw that there was a computer terminal for him to access the network right next to where they were sitting. He went to work trying to research exactly what cell block Jesús was in. He soon discovered that it's highly classified information and not very easy to attain.

After he entered all of their USP credentials to gain access, he finally managed to extract the information. Then they at least knew what cell block Jesús was in. They also learned that he is scheduled for execution in approximately 4 hours. This fact alone meant that they really needed to hustle.

In no time at all the speeding train had reached their destination as it pulled into the station and stopped. When the doors opened a hot gust of wind came blowing in bringing a strong smell like asphalt being laid. The smell had no obvious source other than from the sounds of construction equipment in the distance.

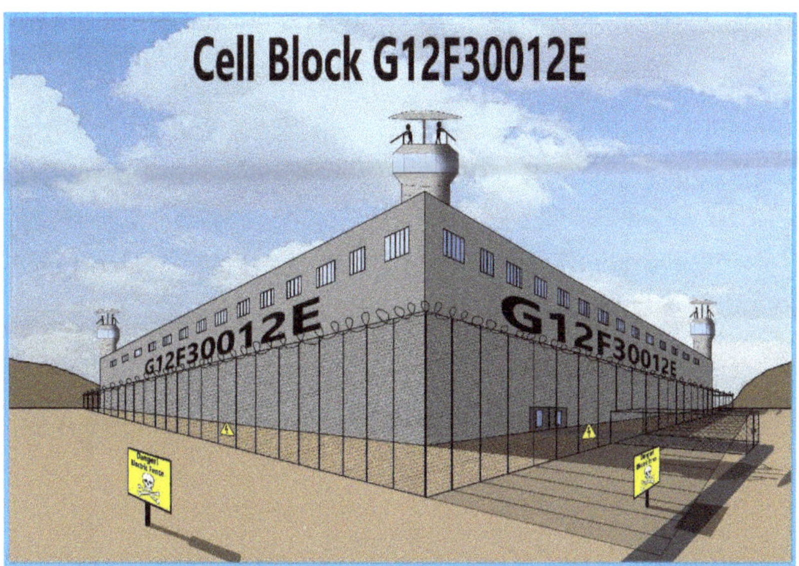

Once they exited the train, right in front of them was the unmistakable cell block building that they were looking for. It was a grey concrete building 4 stories tall, with small, bared windows that wrapped around the top of the walls. There were lookout posts at each corner with armed guards manning the search lights. Most notably painted on the side of the building in large black print was the cell block# GI2F30012E.

As team I walked up closer they couldn't help but notice the 20-foot-tall chain link fence with razor wire along the top just like a prison on Earth, it's amazing how similar the resemblance is. The front entrance was the only semi-welcoming part of this entire building. They walked up to the front door and Abran pulled it open.

They entered the lobby, and they all let out a collective gasp when they saw the spectacular sight. The walls and ceiling were elaborately decorated with panels of beautiful light colored marble that had swirling veins of gold flecks throughout.

There was a service window on the other side of the room about fifty feet away. They followed a long red carpet which had regal looking gold streamers stitched in intricate designs down its entire length. The carpet went from the front door in a straight line all the way to the service window. Next to the service window was a couple rows of very plush and comfortable looking chairs.

Hanging on the walls were paintings of some important looking people, nobody whom they recognized. There was also a few marble statues in each corner of the room. Very notably there was the sound of soft & gentle symphony music playing in the background. Scott was thinking [What the hell! This place is unbelievable! This is far too elegant to be a mere prison lobby. It looks like some grand palace or something like that!] As they walked up to the service window Abran paid close attention to the fact that the lobby appeared to have a low security presence, a possible weak point. Scott was the first one to reach the service window and talk to the security receptionist. The receptionist asked them, "Can I help you?" Scott said, "Yes, we're here for the prisoner transfer #1128." The receptionist said, "Oh yeah we're expecting you guys. Let me see, you know that this prisoner is scheduled for execution in about 4 hours. To hear that he is being transferred might mean that he's either going to have another trial or maybe even pardoned. In any event it's really rare, and almost never happens. Well, I shouldn't say never, I remember last week something similar happened but still, it's incredibly rare. Well anyways I just need for all three of you to positively identify before we can begin the transfer. So, I need each one of you to place your thumb on the thumb print reader for identification, after that we can start the process."

After the brothers had all placed their thumb on the reader. Within seconds all three came back with positive ID's. Once the receptionist saw the results she said, "Well everything checks out! So, we'll start getting the prisoner ready right now. It might take a little while so why don't you just take a seat while you wait."

...**Meanwhile** Team 2 was busy with their mission to disable the tractor beam. They walked into the tractor beam control & administration building. They didn't see anybody and instead of having a front desk, they came upon an automated service kiosk which spoke to them, "How may I be of assistance to you?" Chris said, "We're looking for the tractor beam control room." The automated voice said, "That is located on the 4th floor, room 213. Is there any other assistance needed?" Chris said, "No, that's it." That's when Seth urgently said, "Hey! you guys let's get going the elevator is over here!" Nicole was aware that this building seemed to be completely devoid of any personnel and being 100% automated.

Seth noticed the same thing and wondered if that would make their mission a little easier and faster, because time is of the essence. Seth ran over to the elevator controls and pushed the call button. Right away the doors opened, the elevator was already waiting, they all stepped inside. An automated voice asked them, "What floor do you request?" Chris spoke up saying, "We need the 4th floor." The automated voice said, "As you wish." The doors closed, and the elevator whisked off.

Within seconds, the elevator had reached the 4th floor and stopped. The elevator doors opened to reveal a long hallway that went hundreds of feet to the right & left with just a couple doors either way. Seth stepped out followed by Chris & Nicole. He almost instinctively knew which way to go to.

Seth arrived at door 213, he grabbed the handle and surprisingly it wasn't locked. When he opened the door, they saw a very large room with hundreds of computer banks and what looked like radar and other elaborate equipment. Chris said, "Those must be the tractor beam controls for all the platforms in this area."

There was a platform# on each unit. They started walking down each aisle looking for the controls for platform #327.

Suddenly! they were startled when from around the corner, they heard something metal hit the ground. Seth ran over to see what it was. Standing there was a young man who was dressed in a technician uniform.

It looked like he was trying to pry open one of the control units when the pry bar slipped out of his hands. When the young man saw Seth, he was very startled saying, "Oh my god you scared me, I didn't hear anybody else come in here. Anyways, who are you? And what are you doing here?"

For some reason Seth got the feeling that the young man was hiding something. So, thinking fast Seth changed his facial expression to a serious look and with a stern voice. He said to the young man, "We're here on a monthly inspection. So, I must ask who are you? And what are… you doing here? We didn't see any work request requiring a technician today." Then to be more convincing Seth put the communicator up to his mouth and pretended to call for backup saying, "Copy team I, do you copy? We have a situation here at the tractor beam controls, we'll call you back if we need any assistance." The young man cried out, "Hey wait don't call anybody, please. Look, I don't have a work request to be here, but this machine needs service." Then Seth said, "What kind of service requires a pry bar? I'll tell you what, we won't report this incident. But since you're here we would like some help deactivating the tractor beam for platform 327, do you think you can help us? If you help us, we might be able to help you." The young man replied a little reluctantly saying, "Alright you got a deal, but do you mind if I ask your names." Seth replied ," Alright, I'm Seth and that's Chris & Nicole and we're here on official business. So, what's your name?" The young man said, "My name is Alex….Alex Lextron and I'm here on official business too." (Alex winked at them letting them know that he was aware that they had no business being here either.)

Then Alex said, "I don't know what kind of official business you're talking about, but it can't be too… official otherwise you wouldn't need to have the tractor beam deactivated, that would be done automatically." Then Seth said, "Alright, I must say we're here to rescue our friend from the execution ward and we have come from very far away, all the way from super quadrant 368. Therefore, our mission <u>MUST</u> be successful for a bunch of reasons."

They noticed that Alex had teary eyes when he said, "Your story is very touching to say the least, because it's similar to mine. My mother was accused of smuggling illegal substances across inter-quadrantery boundary lines. She was sentenced to 35 years of hard labor, which is pretty much a death sentence. I swear, I know that she's innocent! She was framed, they set her up to take the fall for somebody else! So, I'm here to rescue her or die trying."

Then Seth said, "I must say that our mission involves three other members, they're actually doing the rescue part and our job is to deactivate the tractor beam. Then depending on how the mission goes, we might need to reprogram the mission profile. So, we can safely leave this place." Then Nicole asked Alex, "How'd you get to this planet, do you have a ship?" Then Alex said, "Yes, I do have a ship. I came here to complete my training as a gravitational attraction energy & radiological technician, in other words I'm a tractor beam service tech." Then Seth said, "Well it's kind of obvious how you can help us, but not so much how we can help you." Alex said, "You're probably right but it would be a miracle if you could figure out a way to get my mom released or even pardoned, like I said it would be a miracle!"

Then Seth said, "Well, you know Alex my friend Jesús is a technological genius and I'm sure that if you help us rescue him. I know that he'll want to help you too." After hearing that Alex said, "OK, guys you convinced me. I will help you because I believe that it's the right thing to do, and I also believe that you will help me and my mom. So, what landing platform do you need deactivated?" Chris said, "It's platform #327." They all followed Alex over to the far wall where the controls for platform #327 was. Alex logged in by waving his card key over a sensor pad on the unit. Once he was logged in, the screen came up with some simple controls. He selected to turn off the tractor beam. Then a prompt asking him to confirm his choice. He then selected yes to complete the operation, then he logged out. Right after that Seth called team 1 on the communicator saying, "Hey guys I've got great news, the tractor beam is off… you copy that!" Then team 1 replied, "That's good news. As far as our mission goes, they're getting Jesús ready for the prisoner transfer, but it'll probably take about 20 more minutes until he's released to us. Once they release him into our custody, we'll be heading straight back to the ship and then we'll be leaving immediately…OVER."

After team 2 heard that they knew they had to act fast. Team 2 replied, "We haven't had time to restructure the return mission yet, but we're working on it…OVER." Then team 1 answered saying, "That's alright just keep working on it, but we may not need that after all. After they release Jesús, we'll know whether or not you'll need to restructure our mission. Until then keep up the good work and we'll talk later…OVER."

After team 2 heard the news they were a little relieved and there was less pressure on them to complete the second part of their mission. Seth noticed that Chris & Nicole were talking to each other in hushed tones, and he wondered what they were talking about. Then everyone turned their attention towards Alex when he spoke saying, "Hey guys, I overheard that you're looking to restructure your mission and for that you would need to go to mission control. It's downstairs, I can take you there."

Chapter 20 _" Escape from the Prison Planet"_

Meanwhile… Back with team 1: The brothers patiently waited for Jesús to be brought to the lobby. Then the door next to the service window opened and out walked three guards with a prisoner whose hands & ankles were securely cuffed.

To describe Jesús; A Hispanic looking man in his mid-forties, about 5'7" with short dark hair. His beard & mustache had some gray in it. He was dressed in the usual prison uniform and had a pair of rimless glasses to complete his look. The guards said, "Well, here he is and be extra careful with this one because he's _really_ smart!" The guards handed Scott the key and then said goodbye! Then they turned and walked back through the door they came out of. Abran was the first one to greet Jesús saying, "Welcome, and we come to greet you in saying you have been selected for a prisoner transfer 1128. If you cooperate everything will go smooth and uneventful." Then Scott said, "We want to put your mind at ease. This transfer is good news for you. We can't disclose all the details right now. Well, anyways we must be going right away." And with that the brothers and their prisoner all walked out the front door heading straight for the train station.

Once they got to the train station, while they were waiting for the train. Scott called team 2 saying, "Hey guys, we got some great news. So, stop whatever you're doing right now! and head back to the ship do you copy that…OVER" Seth replied, "Copy that! We'll see you there…OVER" The minute they ended their communication the train came speeding up and stopped.

The doors opened and out walked a bunch of humans & aliens. Then once it was clear they all walked inside. Then the doors shut, and the train whisked off at over a 100 mph.

…**Meanwhile** team 2 : Alex & them were already downstairs near the administration dept. when they received the news. They stopped everything that they were doing and started heading back to the ship. Alex had overheard the radio conversation and said, "I'm very happy to hear that your friend has been released. And I'm excited that your plan seems to be coming together. Maybe you'll be able to help me free my mother now! With help from your friend Jesús."

Nicole interjected saying, "You know, Me and Chris were thinking about your situation and it's truly heartbreaking. After talking amongst ourselves we've decided to stay here with you and help you free your mother." Seth interrupted saying, "I can't believe what I'm hearing! Are you guys sure about this? Because remember this is a *Prison Planet,* and we might not be able to come back and pick you guys up! So, you could be stuck here for a long time or maybe even forever! Are you absolutely sure that you're ok with this decision?" Then Chris said, "Yes, we're aware of the consequences of our decision, but other than helping Alex we also know how important it is that you guys get off this planet and out of this quadrant as soon as possible. We're aware that the supplies aboard the ship are running low for 7 people, but they will last longer for 5 people. Without us the supplies might last long enough for you to get to the nearest quadrant and restock. And lastly Alex has a ship so we might not be stuck here after all. So, now do you see the logic behind our decision." Seth said," It does make sense, but are you coming back with me to meet up with the others? You can explain your decision to them.

I'm sure they'll understand and accept it too." Then Seth, Chris, Nicole, and Alex all walked back to the boarding platform to wait for team 1 to arrive.

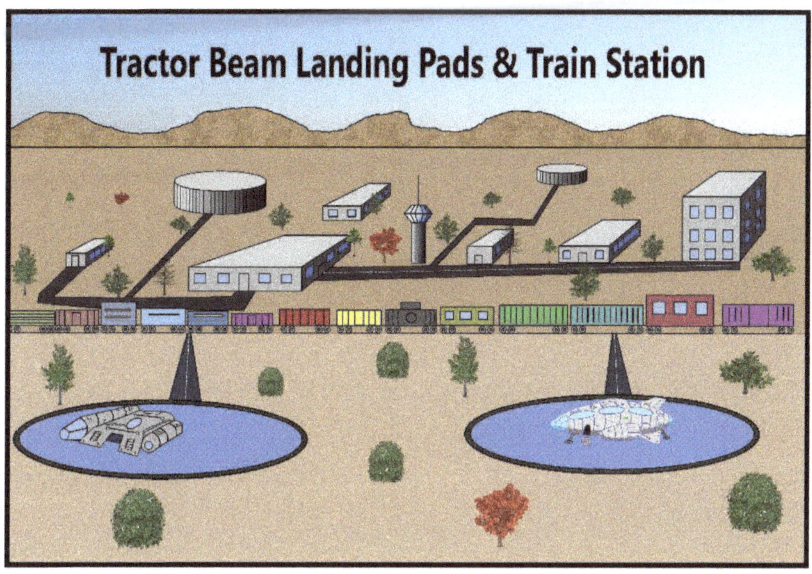

Just as they walked up to the boarding platform the train came speeding into the station. Once the train came to a complete stop. All the doors opened, and a lot of people and aliens came walking out.

Then out stepped team 1 along with their prisoner Jesús. Team 2 walked over to greet them. When Seth saw his old friend, he cried out, "Oh my god Jesús!" Then he went running to greet him. Seth hugged Jesús, who was chained and couldn't hug him back. Seth said, "I'm so glad to see you." Jesús said the same when he saw Seth.

Once Jesús saw Seth he started thinking [Maybe this whole prisoner transfer is just a secret rescue mission otherwise why would Seth be here?]

Scott said, "Well guys as you can see, we're ready to leave so let's get going!" When Abran saw Alex, he immediately became suspicious and said, "Who is this?" As he pointed at Alex. Seth said, "This is Alex, he's our new friend; he came to the Prison Planet for similar reasons, and he helped us deactivate the tractor beam, I trust him." Once Alex was introduced and had shook the brothers' hands, they all said, "It's nice meeting you, and any friend of Seth's is a friend of ours too."

Seth interrupted saying, "Hey guys before we go… There's been a change in plans. Nicole & Chris have decided to stay here with Alex and help him rescue his mother, who was wrongfully imprisoned." Scott inquired to Nicole & Chris, "Is this true?" Both Nicole & Chris affirmed what Seth had said and told the brothers the same reasons that they told Seth. After hearing their explanation Scott said, "Well I guess I have to agree with your reasoning. I just hope that this new endeavor that you guys have chosen is successful for you. I assume that you should be able to hitch a ride off this planet in Alex's ship." Alex said, "I assure you that even though you just barely met me. I'll make sure that we'll take care of each other and get off this planet together. You have my word." Then Scott said, "That makes me feel considerably better knowing that you'll take care of each other. So, in any case I must tell you that this decision is very heart wrenching. Anyways, since we're still on a tight schedule. We must say goodbye…My friends we won't forget you and look forward to seeing you again in the future." Seth chimed in saying, "Don't worry we'll help you just like we said we would. Me and Jesús are already working on a plan right now!"

And with that said, everyone exchanged warm hugs while wishing each other good luck. Then Abran led the way back to the ship and pushed the door button. The door opened and the brothers along with Seth & Jesús walked onto the ramp. Abran went to the controls and started the engines. As the ship slowly lifted off, they all stood at the edge of the ramp and waved a very sad goodbye. Once they had moved about 200 ft. away Abran shouted for everyone to come inside and sit down. Then Erik closed the door and Abran engaged the engines. Then they blasted off, and from the window they watched their friends and the Prison Planet whizzing away at an ever-increasing speed. They headed straight for the nearest space highway on-ramp. Everybody felt sad for leaving Nicole and Chris behind, but somehow, they knew that they'd be ok.

Without being prompted Abran chimed in saying, "I'm gonna miss Nicole & Chris. For some reason I feel like I can trust Alex to help them, and I believe that he has a sincere, genuine heart. So, I know that we'll see Nicole and Chris again someday, I just know it!"

They reached the on-ramp and while dodging all the traffic coming & going. They smoothly & easily merged onto route 3346 heading back through the spider galaxy. Scott gave a warm welcome to Jesús saying, "Listen Jesús, as you can already tell that we're friends with Seth. My name is Scott, and the guy driving the ship is Abran, and Erik is our navigator. We're all brothers originally from Earth and we met Seth through a couple mutual friends Kira Xena Jones and Mr. Bradley, all the way back in super quadrant 368. Anyways when Seth told us your story, we all agreed that we must try to rescue you, so that's exactly what we came to the Prison Planet to do. Seth tells us that you're very, very good with computers and technology, is this correct?" Jesús replied, "Yes." Seth kinda butt in saying, "Don't be modest Jesús, you're an absolute genius. That's why they arrested him on the suspicion of being a life hacker." Then Scott said, "Well in that case do you think that you can help Chris, Nicole, and their friend Alex. From my understanding Alex's mother has been wrongfully accused and imprisoned for the false crime of smuggling. Alex helped us rescue you, so we want to help him too. Therefore, if you think you can help him, please feel free to do so." Erik said," Maybe we should wait until we leave this super quadrant before we do anything?"

Then Jesús said, "It makes no difference of where we are, but if I do something now your friends won't have to wait as long." Jesús walked over to the navigation computer that Erik was sitting in front of. He humbly asked Erik if he could use the computer, Erik moved out of the way. Jesús sat down and immediately went to work. The brothers were pleased and amazed that everything that they had heard about Jesús was true. The man was a natural when it came to computers.

Within a minute or two of searching the Prison Planet data base he had found a woman named Linda Lextron who is incarcerated for smuggling. When he dug a little deeper for more info. He found that her next of kin was named Alex Lextron which confirmed that they had found the right person. He also discovered exactly where she was being held.

Jesús described the process for getting her released saying, "I will have to create a fake pardon. First by using your USP credentials and making a reply for a classified information request. I'll make the request to the regional governor's office who has control over the Prison Planet where Linda is being held. Our request will be for a group of randomly selected prisoners with the sole purpose of copying the gubernatorial transmission. Once they process our request and send the classified transmission back to us. I'll be able to copy the authentication code, from the transmission which is embedded into the signal and is very difficult to extract. So, when I compose our fake pardon and implant the copied authentication code it'll be indistinguishable from a genuine transmission. To hide everything, I'll send it over numerous proxy networks to conceal its origins. When it's done right no one really questions the credentials or the authenticity and if they do, it's nearly impossible to tell the difference."

Jesús started creating a classified information request and sent it. They waited for a response, which took about fifteen minutes. Once they received it, he searched through and dissected the gubernatorial transmission to find the all-important authentication code. Jesús said, "Most people don't know that the embedded authentication code is at the beginning of the transmission. The good thing is that it doesn't need to be unencrypted, all you do is just copy/paste."

Once this crucial element was extracted, he could then compose their fake gubernatorial pardon. The pardon included Linda and about nine other randomly selected prisoners to hide their intent.

Once the transmission was finished and included the authentication code, he sent it. Then Jesús said, "Even though the transmission will get there in less than an hour, it'll still take a week or more for the bureaucratic protocols to run its course. After everything is verified then both the Prison Planet officials and the regional governor's office should accept the fake pardon as legitimate and real. Therefore, you can consider that Linda Lextron will be a free woman after everything is done."

As they sped away from the Prison Planet, they'll have to pass through all the checkpoints that they've already went through. This time they won't be required to verify everything over again, because the highly regulated verification & vetting process has already been done. If they were civilians they would have to do the entire process again. As they left super quadrant 92, they all breathed a sigh of relief.

Chapter 21 " The preponderance of time travel"

As they sped along the space highway everyone began to relax. Seth was the first one to break the silence by telling Jesús, "Man you wouldn't believe what we went through to rescue you. I still can't believe how everything happened so perfectly, it's a freakin' miracle. I'm telling you that it must be pure divine intervention." Then Jesús said, "I can't believe it either. I thought that I was dead, even from the moment they arrested me I knew I was a goner. So, I prayed almost day & night that I would be saved and the minute I saw you Seth, I knew that my prayers had been answered." That's when Scott said, "Well, I hate to remind you that this is a prisoner transfer not a pardon. So, you're not out of the woods yet unless you've got a great idea. Otherwise, we only have a certain amount of time to complete the prisoner transfer. If we're not back to base within a certain amount of time that will trigger a red flag man hunt for all of us." Then Seth said, "Why couldn't we just do one of those gubernatorial pardons like you did for Alex's mom." Jesús replied, "I was thinking about that before I sent the transmission, but I realized how it wouldn't work. For one thing there's already a prisoner transfer logged into the system and that is very difficult to alter. It's too late for the gubernatorial pardons. They don't send those very often, so we'd have to wait for a while." Then Abran said, "Maybe we could go back to Aqualaysha so that Jesús can get a life hack like we did." Seth interjected saying, "That won't work either because since you're USP they'll run your ship's ID# and find that there's a prisoner transfer and that will trigger a red flag. The officials would see right away that something is wrong."

Then Abran said, "Man, I wish we could just turn back time, that way me & my brothers could go back to Earth and save our mom from being killed. And you guys probably wouldn't be in the predicament that you're in now."

After hearing Abran mention time travel, Jesús Started thinking about the feasibility of this suggestion. Then Jesús said, "You know I've heard of an old theory about traveling through time to the past, which requires the almost unlimited power of twin super black holes. There are a couple things that make it incredibly difficult and dangerous. One thing is that you must find them because they are a very rare phenomenon, but we're in luck because there are two known twin super black holes in ultra-quadrant 14."

Erik brought up the map which showed that they're approximately 10 super quadrants away. Then Erik added, "So, with our enhanced speed it would take us about 6.5 days to get there."

As they discussed this preponderance of possible time travel. Seth asked Jesús how safe it was. Jesús said, "I honestly don't know, and I'm not gonna lie to you. It has to be dangerous, anytime that you get close to a black hole of any kind the element of danger is undeniable!" Seth added saying, "You're not actually serious about this time travel thing are you? because it sounds like crazy talk to me!" Then Scott told Seth, "Well given our current circumstances we can't think of any other choice right now. I guess when we decided to rescue Jesús none of us gave it much thought of what to do after. Maybe we kind of painted ourselves into a corner. So, unless you can think of something better? It looks like we're gonna have to try this time travel thing no matter how dangerous it is."

Seth just shrugged and nodded in agreeance. They searched on the info-net and found the time travel theory that Jesús spoke of. After carefully looking it over the group of friends all decided that the reward of time travel was kind of worth the risk. So, they plotted their course towards ultra-quadrant 14.

Secretly Seth had some serious doubts about this time travel thing. He felt scared about the prospects of the whole thing which seemed far too dangerous for his tastes. Seth mentioned, "It seems that this time travel theory is breaking the laws of physics, therefore I doubt it's even possible." Abran said, "Maybe not Seth, just remember that we're defying the laws of physics just by traveling on the space highway."

That's when Erik added in saying, "Yeah, Abran's right there's at least two physical laws that we're breaking right now. One of them is that no massive object can go the speed of light and yet we're traveling millions of times faster than the speed of light right now. Then the other thing is that time slows down the faster you go, and yet everywhere we go time seems to be consistent." Then Scott added saying, "Since the universe is expanding right? Then the space highway must be expanding with it, correct?"

They all looked to Jesús for the answers to these complex questions. Then Jesús took a deep breath and tried to answer these questions the best that he could saying, "As you might know that traveling on the space highway is not like traveling through relatively stationary space, instead the object is stationary but the space that the object is in, is moving at a super accelerated rate carrying the object with it. The source of this warp energy are ultra-blackhole ejection tubes that are focused and directed into special inlets for powering the space highway. At the opposite ends where they terminate are called white holes. It's the way of the universe balancing itself out. By using this form of transportation you're effectively traveling through the 5th dimension. As for the second question, time normally slows down the faster you go, but as I said before we are not traveling through time & space we are being carried by it. The space highway effectively allows space & time to balance itself out around the universe keeping time & space consistent. Finally, the third question is that space is indeed expanding, but in some places, it's actually contracting. So, the more distant an object is, the faster it appears to be moving away which is often an illusion. The movement of space is not even nor is it perfect and the result is that some space has different physical characteristics, which are hard to explain. To answer this question simply, if any 2 points on the space highway were to move either closer together or further apart the space highway will move as well. That's one thing that makes it so ingenious. And that's all I know, I'm not an expert."

After Jesús was done talking everyone was completely flabbergasted. Abran said, "You sure sound like an expert to me." Seth said, "See, I told you guys that he is amazing." Scott said, "Listen Jesús, as you already know that me and my brothers are originally from Earth, but what you don't know is that we're actually fugitives. We got a life hack and in order to complete it we had to become USP. I must agree with Abran about the possibility of going back in time. If we could change the past and rescue our mother from being murdered, sounds incredibly enticing. The vision of seeing our mother being murdered is still to this day an unbearable thought to remember. We are willing to risk the dangers of time travel for that. We also would like the ability to go back to Earth someday. So, I think the one question that we would all like answered is, have you ever heard of any bona fide proof of successful time travel. And if so, is there any paradox from it?"

Jesús responded with, "From what I've heard there is proof, but not much. I've heard of at least 5 cases where time travel was attempted, and the circumstantial evidence appeared that it was successful. This is what the theory says verbatim, (By using this method the physical constraints of time & space are completely removed. Once you are released you can travel through all time in an instant.) This theory really makes you question the very fabric of reality itself. So, from the way the theory is written, if you do everything right, time travel appears to be safe. It doesn't say that it's safe, I'm just making an assumption, but we all know there's so many ways things can go wrong."

As they sped along towards ultra-quadrant 14, in search of the twin super black holes. There was an un-nerving quiet, everyone was lost in thought. Everyone was trying to determine whether they were making a good choice. Due to the fact that this choice is life altering or possibly a suicide mission. Everyone had their own worries about what to expect.

Seth was the only one who decided to himself that he didn't want to go. Not just because it's very dangerous, but mostly because he wanted to stay in the (now) time. Maybe skepticism had got the best of him, but he kept these thoughts to himself.

Until he could figure a way out of this situation, he was kinda feeling obligated at this point.

On the other hand, Jesús realized that there really wasn't any other choice for him. Even after they complete the prisoner transfer that doesn't change the fact that he was convicted of being a life hacker. Even then he'll still have a death sentence hanging over his head. Then there's the far off possibility of getting a life hack which is insanely risky for the person performing the life hack, you'll never find anyone crazy enough to do it. Because the life hack probably wouldn't work anyways. Therefore, time travel might be the only way for some sort of life reset. For Jesús's there simply isn't any other choice, and desperate times call for desperate measures, which means he has nothing left to lose.

After the 1st day of travel, they became aware that they needed more supplies. Erik started searching for a destination where they could get supplies. He found a small outpost located in super quadrant 21 right next to junction 124 located in the whisper galaxy. They should be there in about 1½ days which should leave them plenty of time to get supplies before they'll need to rely on their recycler.

After the 1½ days had elapsed, they were fastly approaching the designated outpost within super quadrant 21. Erik said, "We're approaching junction 124." Abran said, "I wonder what this place looks like, so far we've been almost half way across the known universe and seen a lot of different things. I doubt that we'll be surprised and like usual we should prepare ourselves for the worst."

Then the proximity alert sounded telling them that there was only 30 seconds left until they'll reach the exit. Then the final countdown   10,9,8,7,6,5,4,3,2,1   then they smoothly and seamlessly exited the space highway.

Chapter 22 " Another friend departs"

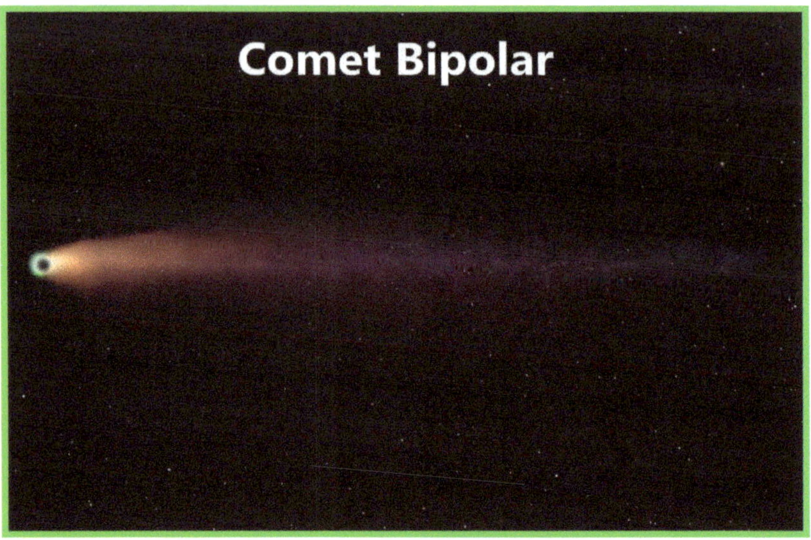

Once they had exited the Space Highway. Erik looked over the map that he had downloaded. It showed that within a mere 5 million miles away from junction 124 was the outpost where they could get supplies. Erik noticed that the outpost was a comet orbiting some sort of malformed solar system. The solar system consisted of a red giant star which was being orbited by a rather extensive asteroid belt. The comet they were looking for was the biggest and the only one that was colonized. The comet's name was Bipolar. It was made of rock and minerals instead of ice like most other comets. Once it was pinpointed on the map, they headed straight for it. They estimated it would take about 35 minutes to get there.

As soon as they got close enough to see more details, they saw that Bipolar had a beautiful, shimmering, orange-red colored tail. The outer bands of the tail had a very faint, light blue color.

There was a few other, much smaller comets not very far away, but they didn't possess the beauty that Bipolar had. As they got even closer Erik calculated the best approach vector. He and Jesús both agreed that it would be best for them to approach from the back and follow up the comet's tail. Surprisingly, from their calculations they should expect less turbulence this way.

The moment finally arrived, Abran followed Erik's approach vector and guided the ship right behind the comet and up through the tail. The ship started to encounter some turbulence caused by the stream of ionized particles that made up the comet's tail. The ride got bumpier the further they went into the comet's tail. It got so bumpy that it felt like the ship was gonna shake apart. Abran shouted, "Something's wrong, we need to turn back!" But Erik said, "No wait! We're almost there." Just when the shaking got to its worst, Abran instinctively reached for the controls to turn the ship around, and then… the shaking stopped! Abran and everyone breathed a sly of relief. Now that everything was calm again, they looked out the window, and saw a bustling, densely packed city full of brightly colored lights.

Once they got close enough to land, they saw that there was plenty of open spaces. The fact that there was no one giving directions about where to land must mean that you choose for yourself. So, that's exactly what they did. Once they landed Erik did a complete analysis of the outside conditions. Like usual he found that the entire surface had breathable air. Therefore, he surmised that there must be some sort of energy shield covering the surface to keep the air in. Which seems obvious, otherwise without the shields this whole city couldn't exist.

The next discovery was that the amount of gravity must be artificially enhanced, because it's only slightly less than Earth's gravity, and is more than expected for this small comet. It served as a base for a major mining operation, mining rare minerals and elements only found on the comets & asteroids in this solar system. After the analysis was complete, they determined that it was perfectly safe to go outside.

Abran walked to the door and pushed the button to open it, and like usual they all grabbed their weapons. The door opened slowly, and Abran was the first one to walk outside. To their amazement it looked like an old western town with high tech accents. The whole city was maybe 75 miles across. The first thing they noticed was the faint smell of electronics and sulfur. The next notable thing was the beautiful, but eerily red sky. It looked like a sunset in an old western cowboy movie and was the result of the charged ions hitting the shields surrounding the comet. They started walking towards town looking for the supply shop, they walked by all the other Ships parked there.

As they were walking Abran noticed something very peculiar. He saw a ship that looked very familiar. So-much-so that Abran said, "That ship looks like Bradley's doesn't it?" Both Scott & Erik agreed with Abran. Then Scott said, "That's impossible, they told us that they were heading back to Marcus 12, and that's halfway across the universe." That's when Seth interjected saying, "Somehow, I just know it's them, I just feel it. I don't know why they're here, but it's fate that they're here!"

Immediately Seth felt a glimmer of hope that his idea of not wanting to time travel, and this might be his only real chance of getting out of this predicament.

The friends slowly and cautiously walked down the busy city streets heading towards the downtown. As they were walking down the streets one thing that was notable was how the scenery changed from an old western town into a very modern and high-tech looking one. The faint smell of electronics and sulfur went away and was replaced with fresh air. For the most part everyone seemed to get around by walking or having small motorized carts similar to go-carts, because large scale transportation wasn't necessary. Along the way Erik revealed that he had brought his communicator, so they could easily locate the supply store. Which was the reason why they came here in the first place. According to the communicator it was a ten-minute walk. As they continued walking the scenery kept changing along with the pedestrians. At first there were mostly humanoids then the further they went into town they saw more and more aliens. The section of town where the supply shop was located looked like a scene right out of blade runner. Scott was thinking [What a contrast between cowboy country and super-high tech.]

Our friends arrived at the supply shop when they walked up to the front doors. The doors opened automatically to reveal a small shop that was very clean and well stocked. They heard a commotion of loud voices coming from the back of the shop. They couldn't see who was talking but it caught their attention. Abran said, "I recognize that voice, I think it sounds like Bradley!"

No one else said anything, but they all felt compelled to find out if it was true. They quickly walked towards the back of the shop.

When they rounded the corner at the end of the aisle, they caught a glimpse of two familiar faces. It was Bradley & Kira!

Who both turned to look and see who had just come around the corner. The instant spark of recognition & joy could be seen on their faces. Everyone was surprised to see each other. After the initial shock wore off Bradley went to greet everyone. He extended his arm in a handshake gesture and Scott reciprocated. Once the handshake began Bradley pulled Scott closer while saying, "Who am I kidding, come over here and give me a hug." As they embraced, they both said, "It's good to see you guys, we're so glad you're safe." After everyone had exchanged warm hugs, firm handshakes, and kind words about missing each other. The next thing Abran asked Bradley & Kira was, what they're doing here? Kira answered, "We're actually heading to super quadrant 92." Erik said, "Really, we just came from there. Why are you guys going there?" Kira replied, "We're looking for my friends Chris & Nicole, and we heard that they had been arrested and taken to the Prison Planet. And you guys should know that sounds bad! So, that's why we're heading there." Seth said, "Oh my god, this is so uncanny because we're the ones who took them to there. But we didn't arrest them, they went with us to help us rescue our friend. To make a long story short, they ended up staying behind to help another friend rescue his mother who was imprisoned there!" Erik asked. "How'd you guys find out about their detainment?"

Bradley told them, "Our friends Ralph & Robyn who are tow transport operators were called to tow a ship back to a USP base. After they checked the ship's log, they found that it belonged to Nicole and Chris. They had been arrested for smuggling and were being taken to the Prison Planet. So, Ralph & Robyn called us to let us know the news. Then they deleted the ship's log and towed their ship back to Marcus 12 instead of the USP base trying to keep everything secret."

Then Scott said, "Well regardless of the uncanny circumstances, we are certainly glad to see you and very happy that you're ok. We were worried if you had made it to your destination. After we split up leaving Aqualaysha I remember thinking about you guys often. As you can see that we completed our plans {Scott winked signifying the life hack} and the next important thing involves this guy." Scott signaled to Jesús, and he walked over. Scott said," Bradley, Kira this is our friend Jesús." After Jesús's formal introduction. Then he shook their hands

Then Scott said, "I must say the reason why we went to the Prison Planet was to rescue him. Erik was able to create a mock prisoner transfer using high level USP credentials, and that is what essentially allowed us to rescue him. And now we have a new problem which is the prisoner transfer. Which has a limited amount of time to be completed, or a man hunt will ensue. What you don't know is that Jesús has been falsely convicted of being a life hacker. So, no matter what he will have to face his eventual execution. That leads to our alternate plan which is the possibility of time travel! This is something that we've all agreed to, and it seems to be our only option. We've worked out a plan to time travel based on a theory which utilizes the power of twin super black holes. That's why we're heading to ultra-quadrant 14 right now."

After hearing this Bradley said, "I think you guys are crazy! You know that it's just a theory and nothing more!" Then Kira said, "So, you guys all feel this way?"

When she looked around, the only one who showed any signs of doubt was Seth. Kira looked directly at Seth and asked him in an abrupt manner saying, "What's wrong Seth, you look concerned?" Seth looked around and answered, "Hey guys, I haven't told you, but I've been thinking about this time travel thing, and I don't want to go. I want to stay here with Bradley & Kira. I hope you guys aren't mad. Plus, I want to stay in this time. I don't have any regrets about my past and I don't feel the need to change anything. I don't think it's worth the risk because it sounds really dangerous. You guys should rethink your decision as well!" Scott said, "I kinda had a feeling that you would want to descent from this mission. I must say that I don't blame you, your situation is different than ours. So, I understand & agree with your decision. So, with that said we still need to get supplies to either complete the time travel or the prisoner transfer. Either way if we don't report back to base within 36 hours a manhunt will begin." Erik said, "That definitely means that we can't stay here for long." Our friends all walked together and shopped while reminiscing about the adventures and fun they had together.

Once the brothers had purchased all the supplies they needed, everyone exited the shop. They all went to find a place to sit down and talk. It was an unspoken fact, but everyone knew that once they split up and left this comet. It's almost certain that they'll never see each other again, which created a very somber mood. The area that they decided to sit, and talk had the most beautiful overhead view of the pitch-black sky that was speckled with stars of varying red and blue hues that twinkled ever so slightly. The ionically charged particles that made up the comet's tail created the most beautiful aurora borealis of shimmering colors they'd ever seen. This was coupled with an irresistibly good smelling aroma coming from a nearby restaurant. To complete the mood the air was perfectly warm with a light, pleasant tingling sensation on their skin which felt like a lover's touch. This sensation was caused by the charged particles hitting against the shields surrounding the city electrifying the air, everyone seemed to enjoy this phenomenon.

The nuances of this moment were so captivating that their somber mood was somewhat diminished and replaced by a feeling of happiness. Everyone realized that this could be their last time together as a group, so they wanted to make the most of this special moment. Once they all sat down and relaxed for a bit.

Abran & Bradley volunteered to get some food and drinks for everyone at the nearby restaurant. Bradley asked everyone what they wanted; he had a special talent for remembering stuff like this. Once everyone had stated what they wanted Abran & Bradley stood up and walked over to the restaurant, which was only about 200ft away.

After a short time, Abran & Bradley came walking back with their hands full. They gave everyone the items that they had ordered and then sat down. Being curious Bradley asked, "So, Jesús how far back in time do you need to go in order to fix everything?" Jesús replied, "I would probably need to go back about a year. Then maybe I can rectify most of the unfortunate events that have occurred. I know anything can happen, it might be all good or bad, but if I stay here in this time it pretty much means certain death. Therefore, my choice seems very clear."

Then Bradley asked the brothers the same question and Abran is the one who answered saying, "Just like Jesús said, we know it's dangerous and scary, but having the chance at saving our mother from being murdered is one chance that we're willing to take. Then maybe we'll be able to fix things on Earth, so we can go back there some day."

Then Erik stated, "I've been running our proposed trip through the computer, and it shows that the chances of success are somewhere between 40%-50% which isn't really that good."

Then Scott said, "Well, I know that our proposed trip has some real dangers along with some genuine concerns, but the reward is worth the risk….. I think."

Then Scott commented about Kira & Bradley's mission saying ,"Anyways, as far as your mission goes you guys said that you're on your way to rescue Chris & Nicole. In that case, I've got some good news for you. Before we left super quadrant 92, Jesús created a fake gubernatorial pardon for Alex's mother. By now I'm sure that the Prison Planet officials have released her. Alex agreed to take Chris & Nicole with him when he leaves. I know that you guys don't know Alex, but he did help us, and I honestly trust him to take care of Chris & Nicole. If the plan works then they could be halfway through the quadrant by now. Then you might just have to meet them halfway." Then Kira said, "You're right, that is good news, and it might just turn our rescue mission into a rendezvous instead, Thanks a lot you guys!"

After about an hour of just sitting and talking and reminiscing about the good times. They all sat back relaxing and enjoying the beauty of the scenery and the charm of the moment. When the conversation seemed to stop Scott broke into the silence saying, "Well……it looks like it's time to get moving." Abran was the first one to stand up saying," Oh…I almost forgot we need to fuel for the ship. I'll go now and start doing that. So, I'll meet you guys back there."

Then Abran started to walk off and Bradley stood up saying, "Wait, Abran I'll go with you and show you exactly how to get fuel, it's a little tricky here." Abran stopped and waited for Bradley, then they both walked off together.

Scott, Erik, Seth, Kira, and Jesús continued to talk about things for a while. Then after looking at the clock Erik interrupted saying, "Hey Scott, I hate to interrupt, but I think we need to get going pretty soon." A noticeable wave of sadness seemed to come over everyone for a number of reasons. First is the almost certain fact of never seeing each other ever again. And the other reason is that the brothers are embarking on a very dangerous and uncertain mission.

Also, everyone was having such a good time enjoying each other's company. That was the other reason why they wanted this moment to last forever.

Narrator: *Everyone has experienced a moment in their life when everything is perfect with you and the universe. These moments are few and far between. We all wish that those moments could last forever, and that we could relive them over and over again. Furthermore, every moment matters more than you think. Remembering a joyful time with people who are no longer around anymore. Some people are only in your life for a very short time, then you never see them again. Other people are part of your life, for your entire life. Then some things happen that are unwanted or even terrible. Then later it turns out to be one of the best things ever. Is this just coincidence or God's hand at work, allowing it to happen at all! The past is everything that was meant to be, that was made into reality. The future is created from consequences of the past, and all things come around,* *someday you'll reap what you sow. And with that said we rejoin our friends…*

Scott stood up then so did Erik then everyone did, then they started slowly walking back to the ship. Along the way Scott whispered to Jesús about how worried he was after hearing the low percentage of success that their mission had. This revelation was profound, but the reward of success slightly outweighed the risks.

As they turned the corner, they saw the ship along with Abran and Bradley, both standing under the ship fueling it. Everyone met up right by the ship. Bradley was the first one to say, "Well guys I can't believe that this moment is finally here."

Once again, an intense wave of sadness came over everyone. Then they realized that there's no more time to waste, it's time to go! Everyone started to say their goodbyes and hug each other.

Kira gave each one of the brothers a small kiss on the cheek. Then she said, "I still remember the day we met, or should I say the day that I was captured by you guys. You know, even from the start I didn't feel like I was being captured, it felt more like I was being recruited. To tell you the truth before you guys showed up my life was so incredibly boring to the point where I almost felt depressed. I felt a yearning for a change in my life. When I saw you guys, I just had a gut feeling that the moment for change had arrived. I'm glad to have met you guys and regret to see you go."

Then Bradley shook everyone's hand while saying, "Man, if it wasn't for you guys, I probably would've never been reunited with Kira and got my ship back. Soon we'll be reuniting with Chris and Nicole, and then with Seth tagging along with us we'll have the whole gang back together again, and that's all because of your help. So, now I sincerely wish you guys an absolutely safe and meaningful trip. Goodbye my friends!"

And with that said, Bradley walked over to the ship to check on the fueling machine which looked like it was done. He stopped and disconnected it. Erik walked over to the brother's ship and pushed the button to open the door. The door slowly lowered to the ground. Once it was all the way down Abran pushed the supply cart up the ramp into the ship. Then Abran, Scott, Erik, & Jesús all walked into the ship.

Once everyone was inside Erik pushed the button to close the ramp as the door was closing, the friends all waved to each other until the door was completely shut. Abran & Scott went straight to the ship's controls and started the engines.

The ship started to slowly lift off and as soon as they were high enough, they took off heading towards the tail of the comet to exit. Like before once they got to a certain point in the ionosphere, they started to experience the same ruff turbulence that they felt before which was pretty bad penetrating the shields and entering into the stream of particles. The sudden jerk of the transition was kinda violent but stopped very quickly after they fully exited the comet's tail. Once they exited, they headed straight to the nearest space highway on-ramp.

Chapter 23 " Into the unknown"

Once they had reached junction 124. Erik announced that they had to take route 999 and that should take them all the way to ultra-quadrant 14. They performed the same procedure that they did before to get back to their maximum speed and within about 20 minutes they had accomplished that. Once they were able to obtain all the speed that they had lost due to their stop on Bipolar, their estimated travel time would be about 4.5 days. As they were racing along towards their destination Erik was the first one to open up the conversation by asking Jesús a little bit more about himself.

Jesús revealed that he was born and raised in super quadrant 369 in the Nebulionic galaxy on a planet called Leptron. Like most inhabited planets it's similar to Earth when it comes to all of the life sustaining ecological conditions. However, one of the main differences between Earth and Leptron is that it was terraformed about 500 years before Jesús's family migrated there. Everyone on the planet Leptron is still dedicated to continuing the terraforming process. Sadly, Jesús's grandfather was orphaned when he was very young. Therefore, he doesn't know very much about his ancestors. Most of the records were lost or destroyed during the transition stage. Jesús's childhood was very good with loving parents who helped foster his noticeable talent for technology. Even at a young age he was a prodigy and was able to figure out even the most perplexing problems that he was presented with. As he grew into manhood, he yearned to expand his knowledge. His home planet was sparsely populated with very limited resources.

Therefore, he knew that he would have to leave his home planet and venture out into space to find new horizons. Through credible sources he found that some of the best schools in the known universe were in super quadrant 368. Jesús did some research and found a galaxy close to the boundary line between the two quadrants. In the Micro galaxy he found the planet Digitarus which has a school where he could get the education he was seeking. Jesús left his homestead and traveled to the nearest township 40 miles away to seek out a transport. The nearby township was named Denak which only had a population of 3000, but most importantly it had a spaceport.

He asked a number of people, and someone referred him to the local info-net café where the barista who works there pretty much knows everybody coming and going. He entered the café and asked the barista about needing space transport. The barista pointed towards a dimly lit corner in the back of the café and said, "Do you see that guy sitting way back there, I think he's the person that you should talk to. His name is Seth Phenix. I believe that he has a daily route between here and super quadrant 368, therefore he's probably your best bet." Seth worked as a courier bringing supplies back and forth. Since the planet Leptron has a small population it doesn't need a huge logistics system to deliver supplies, just a simple courier will do.

Jesús walked back to the dimly lit corner of the café to go meet Seth. When he got there, He humbly introduced himself and told Seth about his dilemma. They sat together and talked for a while and got to know each other. They seemed to hit it off right from the start.

They joked and laughed about things, and it was an instant friendship. Jesús told Seth that he had found a planet close to the boundary line between the two quadrants where he would like to go. Seth agreed to take him, and it was only a small deviation from his regular route. Once they made the plans, they set out to go there.

After Jesús had pretty much told the brothers all about his past. Then the next focus was on getting to ultra-quadrant 14 as quickly as possible. Erik & Jesús started searching for the time travel theory that they planned to use, which involves the two super black holes. Just like all things in the universe, black holes are spinning and have polarity, which means timing is essential.

Once they found the theory and downloaded it, they were able to calculate their approach vector, along with the speed & timing. They were able to calculate the exact moment when to start their descent between the super black holes. After they analyzed the theory they began to feel more confident that this mission might be successful. The only part of this theory that was the most concerning was the fact that it was unclear if the journey through the time dilation vortex was survivable. Found within the theory was an important phrase which said { *Your consciousness is the only thing that is truly capable of traveling through space & time.* } Then the next logical question is… What about the rest of you?

After what felt like the longest 4 days of their lives they finally reached the borderline to cross into ultra-quadrant 14. Erik had set the proximity alert to go off when they crossed.

Then only moments later the final nerve racking seconds had come before crossing the borderline. Within 5,4,3,2,1... Now they had entered into ultra-quadrant 14.

Within about an hour-or-so of traveling they noticed that some of their instruments started to act a little funny. They assumed that it had something to do with the intense gravity in the area. Even though they were still a vast distance away from the super black holes their gravitational influence could still be felt. Erik pulled up the map to ultra-quadrant 14 on the computer to thoroughly plan every moment of their journey. He estimated the distance to be approximately 5.7 million light years until they'll reach the nearest significant object that will cross their path, which is the exodus asteroid belt. They shouldn't experience any slowdown while passing this object.

Over the next 3.7 hours they continually looked over the map carefully re-calculating their approach vector as more information became available.

Jesús noticed that beginning somewhere around 17.4 million light years into this quadrant that the super black holes have a spherical zone around them that encompasses the entire area. Within this zone an anomaly occurs, the space highway mysteriously starts to dissolve, and then it rematerializes on the other side. It seems obvious that the dissolution of this section of the space highway is caused by the intense gravitational turbulence in the area. This was a startling fact that they didn't expect to find. Without the speed of space highway, they will only be able to maintain a maximum speed of 549 million mph. Although this speed may sound fast it's nothing next to the mind-blowing speed of the space highway.

This new information created a major dilemma, with the prospect of taking too long to reach their destination. Regardless of this fact they realized that they had to start slowing down to reduce their enhanced speed and return to the neutral speed of the space highway. In order to have better control of their speed as they neared the end of the safe approach zone.

After about a half an hour the proximity alert sounded telling them they were now passing over the exodus asteroid belt. Which is stunningly large covering a vast region of space and is more than 2 million light years across. Unfortunately, the only way it could be seen was from long distant images found on the info-net. It is illuminated from the distant light being emitted from the intensely bright excretion disk surrounding the super black holes. As they devour the nearby planets and stars within the surrounding nebula, which provide them with a lot of food. Once they pass the asteroid belt, there is only 7.1 million light years until they'll reach the end of the safe approach zone. After which, the space highway will start to slowly dissolve. Both Erik & Jesús calculated that even though they'll lose most of their speed once the space highway dissolves. They should still retain enough speed to reach their destination in a timely fashion. Jesús made a special mention about crossing the safe approach zone, saying something that was kinda obvious. He said, "Once we cross that boundary line the gravitational intensity really increases all the way until we reach the time dilation vortex. This will greatly affect how difficult it will be to control our speed when we're subjected to these unbelievable forces. Another factor to consider is that the density of space will probably fluctuate wildly & unpredictably. But, perhaps the strangest thing of all is that the space highway seems to be intentionally directed right between the super black holes, as if on purpose. Then lastly the super black holes are indeed rotating around each other as if tied together at a central tether point while spinning in unison. Luckily, this combined motion is slow and predictable. So, we should be able to precisely time our descent right down into the center of the time dilation vortex."

They enjoyed looking at pictures of the asteroid belt on the info-net which were taken from a tremendous distance away. Scott spoke for everyone when he commented saying, "This asteroid belt is really beautiful, it looks like Saturn's rings but on an unimaginable size & scale. Therefore, its immense size only adds to its majestic beauty." After the two hours rolled by, they had completely passed over the asteroid belt and now only had 7 more hours until they'll pass over the safe approach zone. After they tabulated all of their calculations, they determined that their loss of speed wasn't going to be as important as they originally thought. Because by the time they experience any significant slowdown they will be entering the black hole spherical zone of influence. That's when the black holes will displace enough space that they will free fall towards the time dilation vortex at an ever-increasing speed. So, their current speed may be adequate after all.

The area of space that they were currently in was relatively clear and calm. During this time there was a wave of anxiety that swept over everyone. The worry on everyone's face was noticeable, realizing the fact that this seemed to look more like a suicide mission than anything else.

Erik made an announcement that there was only 15 minutes until the prisoner transfer was due to be completed. After its incompletion, which was sure to invoke a full-scale man hunt. By now our friends are too involved in their current mission to worry about anything else.

The hours rolled by without any adverse developments, except that the tensions kept rising. They found themselves only about 25 minutes away from crossing the safe approach zone, which is also the point of no return. Once they cross this point, they will not be able to just simply turn around and go back. Most likely they'll be unable to escape gravity and be forced to continue on. After reading more information about the super black holes. One important detail is that unlike planetary gravity, the gravity being emitted from the super black holes' alternates between positive and negative. This phenomenon is the only characteristic which could render their main engines completely useless. Then the only propulsion they'll have at that point is their conventional rocket thrusters. This new information made them feel very uneasy about having very limited control of their descent into the time dilation vortex. Everyone seriously started thinking about turning around at this point. However, Jesús was very confident that he had an idea on how to modify the function of their main engines to overcome this limitation. After hearing this and having faith that his solution would work and would allow them to regain complete control of their main engines. This made everyone feel a lot better, so they all reluctantly decided to proceed onward.

Then everyone jolted when the proximity alert sounded, once again and maybe for the last time. There was only a minute left until they'll cross over the safe approach zone and into the truly unknown. Then finally the 10,9,8,7,6,5,4,3,2,1… **then they crossed.**

Chapter 24 " The point of no return"

Within just a few minutes they started to notice some subtle changes. One of the changes that they noticed right away is that the computer started giving a different read out for relative velocity. After looking at the map both Erik & Jesús concluded that the space highway is just starting to dissolve and will do so over the next 3.9 million light years. After it completely dissolves, they will have to rely on their ship's engines and the black holes gravity to get them the rest of the way. Once they have slowed down, it might take weeks or even a month before they'll start falling towards the super blackholes. Then after gravity takes over they'll have a different problem to worry about, which is gaining too much speed.

After being quiet for some time Abran spoke up asking a great question saying, "When we get closer won't we get caught up in the tidal forces? And at that point how will we maintain control of our ship?" Jesús tried to answer this question to the best of his knowledge. Jesús answered saying, "Even though the theory may have flaws or might be just plain wrong. In simple terms the answer is Yes undoubtedly there will be tidal forces. But the thing to remember is that there are two black holes of equal mass spinning in opposite directions close to each other. This should cause any matter to be forced down the middle between them. Which should effectively cancel out the majority of the tidal forces that we'll be subjected to. We are relying on this phenomenon to be present in order to be able to navigate between the super black holes. Without this phenomenon occurring our mission is just not possible at that point. So, all we can do now is **pray!!!**"

To pass the remaining time everyone did so in their own way. Abran remained sitting in front of the ship's controls. Even though he was physically sitting there, mentally he was reliving fond memories of his childhood. The precious times spent with his mother, father and two brothers. This was a time of his life that he missed very badly almost to the point of depression.

Erik & Jesús spent their time searching for old video games to play. They found themselves laughing at how difficult those old video games were.

Scott on the other hand was a little more reserved than usual. He left the bridge and went to the rec-area to be alone. He would often quietly talk to himself about important things; Not because he's crazy, but because verbalizing his ideas helped him think. Scott has a bad feeling about this mission and realizes that it might have been a bad choice after all. The only comfort that he feels about their choice is the fact that there isn't any other choice right now! So, they must do everything they can to make it successful, because their lives depend on it.

In no time at all the 3 hour stretch was coming to an end. After a few minutes Erik announced that, "The space highway has now completely dissolved. It'll probably take a few weeks for us to decelerate back to our regular speed. Then we will only be traveling around 549 million miles per hour, which is only 81% of light speed." Abran was the only one who had been monitoring the dissolution of the space highway. He was watching to see how it affected their deceleration rate. After about 12 hours had passed he became somewhat puzzled when he noticed that the measurements didn't match their calculations.

By now there should have been a small, but notably loss of speed, but there wasn't. So, Abran called out to the others to let them know that their speed didn't seem to be diminishing. Once Abran showed them this new fact, everyone was perplexed to hear of this news. They figured that their instruments were not giving an accurate read out. Immediately Jesús commenced a diagnostic on their instruments to verify that the readings were accurate and found that they were in perfect working order. Therefore, the measurements were true. Even though the space highway had completely dissolved they maintained almost all of their speed.

This was unexpected and concerning due to the fact that the theory did not account for this. They figured that there could be more variables that the theory didn't predict, and that was the most troubling fact of all.

When Scott heard the news, he said "I remember hearing a theory that might account for this, which is {The density of space and differential physical phenomenon theory} Which essentially means that some regions of the universe have slightly different physical characteristics due to the fact that space is not perfectly distributed throughout the universe. Therefore, different areas will have different densities of space."

Nevertheless, by retaining their speed meant that they'll still be able to reach their destination within a reasonable time.

Meanwhile... Jesús and Erik were reconfiguring the ship's engines to harness the alternating positive & negative gravity that will be found closer to the black holes. Even though speed isn't an issue right now they still need their main engines to control their descent into the time dilation vortex.

They also have to modify their shields to extend the full power range. Without the *relative* safety of the space highway, they run a much greater risk of colliding with objects in space. Abran realized that their instruments were giving readings similar to being on the space highway. So, the next conclusion was that even though they were still millions of light years away from the two super black holes, their influence was still profound and that may be the only reason why their speed wasn't diminishing.

After another 2 hours went by. Erik made another observation that their speed seemed to be slightly increasing which was obviously the result of getting closer to the super black holes. The other thing they noticed was that their instruments were becoming less reliable. The sector of space that they were entering into was becoming more, and more turbulent.

Abran urgently mentioned that they need to slow down a little. Because at their current speed even something a 100 billion miles in front of them only gives them a second to react and make corrections.

The surrounding space was really starting to densify causing themselves and all the objects around them to fall faster & faster towards the super black holes. As predicted the two opposing tidal forces canceled each other out effectively creating a neutral zone that almost guided them right into the center of the time dilation vortex.

They had reached the 1 million light year mark; and now it was time to get the final preparations under way. One perplexing thing that the theory said was [Your consciousness is the main determining factor that gives you the ability to time travel. Without it, you will not be able to do so.]

They figured out how to regain control of their main engine by pulsing it with alternating positive & negative gravity. This worked, but the closer they got the harder it became to control their speed.

When they reached the 6 light year mark, they had little to no control of their speed. They found themselves free falling into the time dilation vortex uncontrollably, and this is when panic started to set in. Abran was the first one to cry out saying, "Oh my god we're gonna die!" Erik said, "Calm down, we're gonna be ok!" {Everyone kept quiet, but deep down everyone knew how much danger they were in, and with the small possibility of success.}

Chapter 25 " A conscience decision"

Within what felt like minutes, they had reached the 5 light year mark. They could feel the intense gravity acting on the ship even with the shields at full power it was barely enough! They could hear the ship's walls starting to bend and flex under the tremendous strain. The ship's hull alert would sound whenever the walls started to bend. They couldn't determine how fast they were traveling because all the space around them was moving in a turbulent fashion and there is no relative point to measure against. All they knew is that they were traveling faster than their ship was designed to travel. Within minutes they already reached the 4 light year mark, then 3 then 2 and 1 …They couldn't determine their speed, but their approach vector seemed to be perfect according to their instruments. So far, the ship and the shields were holding together, but under terrific strain. According to the instruments they were only moments away from entering into the time dilation vortex 5…4…3…2…1…… then……

Once they entered into it……. Time seemed to slow to a stop. As Scott looked around the ship in a blurred vision, he saw that everyone was still sitting in their seat. That's when he started having flashbacks remembering his whole life being played back, like on an old film projector. Remembering his warm, water-colored memories of his mother & father, grandparents, brothers, sister, and close friends. His flashbacks were of his earliest childhood memories all the way through his young adulthood and then to present day. During his flashback he felt a wave of joy and excitement, followed by fear & sadness, and lastly intense anger.

He experienced all the vivid emotions that somebody would experience in a lifetime in what felt like a few moments.

Then the flashbacks ended as suddenly as they began. Even though the flashbacks had stopped, he still felt like he was in some sort of translucent dream. When he looked around the ship, he noticed that everyone else seemed to be experiencing the same thing. Scott started thinking [Are we time traveling right now? Is this how it feels? Cause it doesn't seem right. What if we did something wrong and that might mean that the worst is happening right now.]

Time seemed to slow to a complete stop! Then all the sound became muffled until everything went completely silent. The silence was broken by a continuous alert telling them that shields were beginning to **fail.** With this failure holes started developing in the shields causing the hull alert to sound telling them **DANGER HULL BREACH!!**

The shields were failing, and the ship was in danger of breaking up!

During this translucent dream like state, they could smell burning plastic and metal. This was followed by the smell of their own flesh starting to burn, and at that moment they all knew that they were going to DIE!! Even this fact was difficult to discern between reality and a dream. That's when they all really understood what the theory meant when it said [*Your consciousness is the only thing that is truly capable of traveling through space & time.*] The theory basically means that within the space between the black holes their consciousness will be freed from their bodies and be able to time travel. Therefore, they will still be conscience enough to complete their mission.

"Once your consciousness/spirit is freed from its physical bonds you can time travel anywhere in universe instantaneously. Time becomes an open canvas; all of time can be seen at once. Therefore, you can think of death as truly being free and gaining God like powers. Unfortunately, the transition between life and death inevitably has variable amounts of pain depending on how you die. No one really knows how painful this transition is, but every situation is unique. Just like birth is painful…So is death."

Narrator:

At this point in this story, I realize that once again in order to continue to effectively tell this story any further. I have to return to a first person perspective; I hope that it's not too confusing to anyone reading this and apologize for confusing anyone.

I, being Scott once again.

 And now back to the ship…..

As I sat in my chair

At this point accepting that our situation is grave, and our death seems imminent. The ship started to feel hot to the point that my skin started to sting a little, then it started to burn quite painfully. At that point I started having the most vivid hallucination I've ever had, and I started having an out-of-body experience. I remember standing up and then I started to float when I looked back and saw myself still sitting there. Then when I looked around, I saw my brothers and Jesús experiencing the same thing. The next thing I remember is, going through the walls of the ship and through the failing shields like they weren't even there. Then I looked back and saw our ship from a far and I was moving further away from it.

I watched our ship with us still inside as the shields completely failed and within less than a second our beloved ship exploded and atomized without leaving a trace!

When I saw that I felt a little sad cause I knew that we had just died, our physical bodies had died!

I looked around to see the intensely bright light and ferocious energy within the super black holes and I was amused by this. The next thing that I realized was that my consciousness was expanding exponentially. And I knew it was true that once you no longer have a physical body, you will free up your mind. All the extraneous thoughts that are occupied pertaining to your bodies physical needs are freed up. Not to mention the fact that your physical body is trapped in a certain time & place. Whereas your consciousness/spirit can be almost anywhere instantaneously, time & space become irrelevant. One of the consequences of not having a physical body is that you lose your humanity not only the good parts, but also the bad.

Once I really understood what just happened to us. I knew that we had to act quickly because after a short time our consciousness will dissipate and will eventually join with the collective consciousness of the universe. Then we'll essentially become a part of God or the force or whatever you want to call it. Having a physical body keeps your consciousness anchored to something and separate from the collective consciousness.

Narrator:

The question is: Why did they have to go through the super blackholes just to die to be able to time travel.

How come they couldn't have just killed themselves instead to achieve the same results?

The answer is due to the extreme power of the super blackholes central vortex. Which opens up a gap in space and time creating a massive rift allowing their consciousness/spirit to exit their physical bodies enabling them to have control over their destiny.

Chapter 26 " Free at last!"

We were free at last! From our physical bodies and still conscience enough to know what we're supposed to do.

I remember…I remember…… what I was supposed to do.

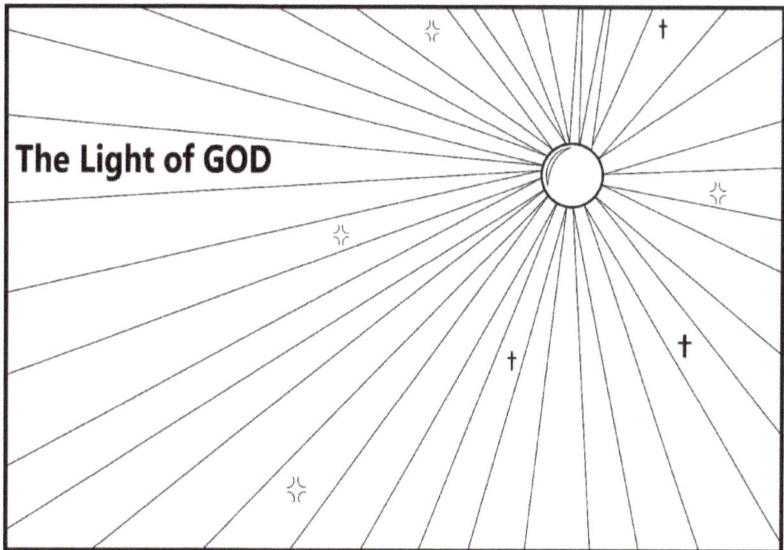

As I moved further away I was able to see the two super blackholes from a far. I kept moving further and further away until I realized that I had left the universe and couldn't see anything. Then suddenly I felt an **immense** presence behind me. I turned to look and saw a warm and extremely bright light coming towards me. Even though it was <u>so</u> incredibly bright it didn't hurt my eyes to look at it.

That's when I realized that I didn't have eyes. The light was absolutely the most beautiful thing that I had ever seen and felt.

I knew right away that this was… The light of God. I could feel the boundless love emanating from it.

The light felt so good and warm. I felt myself willingly & uncontrollably going towards it. As soon as I was ready to enter the light with absolute euphoria! Something told me that I wasn't supposed to go into it.

Suddenly I felt something pull me away and back into the universe. Then instantly the light was gone!!! At that moment I felt **incredibly sad** knowing that I wasn't going to enter the light.

When I looked around and I could see the entire dimensional landscape of reality. Which looked like an infinite number of encapsulated universes containing their own realities.

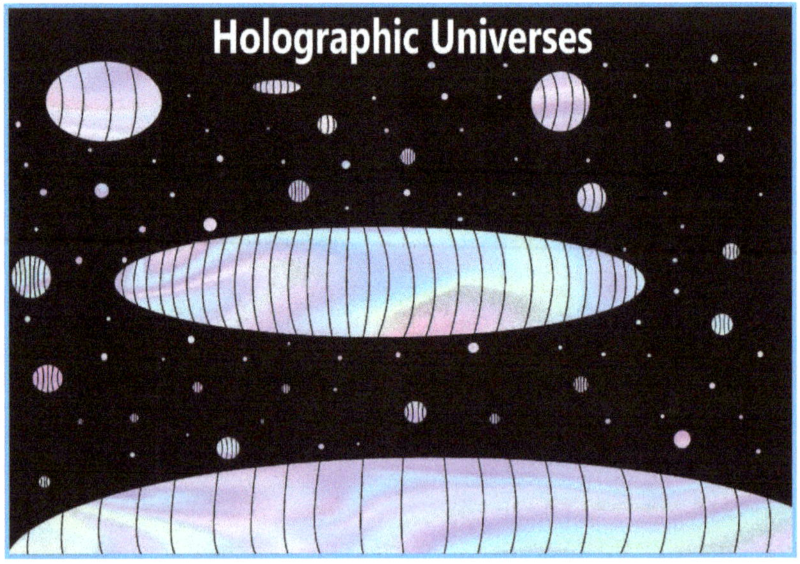

The only way I could describe this in a human sense is that it looked like some sort of panoramic hologram. This unbelievable ability of being able to see the entire holographic reality wasn't like seeing something with human eyes.

Instead, it was being able to *perceive it* , and fully comprehend reality without the limitations of the human mind.

I was also able to conceive of the concept of absolute nothingness. Which is [No matter, no space, no time, and not even God!] Absolutely nothing, which is actually an extremely scary thought!

Then I felt something pull me again and within an instant I found myself back on Earth right above our old house.

That warm & great feeling of **HOME** came over me like a wave of goose bumps. Then I entered the house and saw my mother alive! and I knew it was her! The time travel seemed to have worked!

I felt another pull and this one felt the strongest so far. I was pulled out of the house and straight to the underground laboratory. Where I saw myself, and that feeling of **HOME** was <u>so</u> profound. I remembered what I was supposed to do. Then I was pulled one last time, back into my body.

Chapter 27 "Rejoin Body and Soul"

Narrator:

Their consciousness otherwise known as their spirits have transcended time & space to rejoin with their living selves in the past. Once they're rejoined they will have very subtle and indirect forms of communication. Their spirits will be able to influence them with a strong sense of intuition, otherwise known as a gut feeling, to help to guide them. This is Similar to being protected by a guardian angel.

and now back to the living...

While I was sitting in my chair I suddenly experienced an epiphany. I had an overwhelming prophetic vision, a sense of intuition that I just couldn't ignore. I felt a strong feeling that we need to get ready to leave the laboratory and something is approaching that we should flee from.

I turned to my brothers and said, "Hey guys, this is gonna sound weird, but I just got this feeling that something is coming, and we need to get out of here. We also need to tell mom because she's in danger too. She'll need to come with us!"

My brothers told me that they experienced the same prophetic vision that I did. Which meant that this vision was something truly special and absolutely couldn't be ignored. After which we all expressed the same level of urgency. Erik said, "I wonder how mom will react when she sees our ship & laboratory?" Then Abran said, "It doesn't matter. I'll go get her; you guys keep getting things ready so we can take off as soon as possible." Then Abran walked back to the house to get mom.

I started experiencing more strange visions while me and Erik were preparing to go. These visions felt like déjà vu or some sort of daydream. Featuring us first being chased off the Earth by some governmental agency. Then we met some strangers in space who became our close friends. Then we traveled vast distances at incredible speeds on an inter-universal highway system. Then we traveled to a water planet where we met another close friend who helped us get a life hack. Then after that we became United Space Police. Then we helped rescue another close friend from a Prison Planet. Then we were briefly reunited with our close friends while we were traveling on a comet. Then finally we traveled between two super blackholes in the attempt to time travel, but we tragically died while doing so. I felt that these visions were more than just a daydream. I somehow knew that they were a prophetic vision of possible future events. I almost disregarded them as nothing, but they were so incredibly real that I couldn't ignore them.

I kept quiet and didn't tell my brothers, but these visions stuck in the back of my mind.

Nevertheless, we continued to get the ship ready to go. Abran came walking back with mom, she asked us what's happening. She was shocked to see everything, the laboratory, our ship. She was almost too stunned for words. Then suddenly the outside motion detectors went off picking up movement. So, Abran switched on the visual display to see a whole bunch of cars & vans pull up to the front of our house. He noticed that all the vehicles were black and un-marked. As soon as they stopped, then about 50 or more agents got out carrying rifles and machine guns.

Once they had completely surrounded the house, then they ran up and kicked down the front & back doors. The agents stormed all through the house searching for anybody. They searched everywhere and couldn't find anybody. Then they brought in an armored personnel carrier with a battering ram on it.

We watched all this from the relative safety of our laboratory, and said to each other, "We have to launch right NOW!" We had planned for this type of emergency, in case the police, FBI, CIA or any other governmental agency was to ever raid our property looking for us. We were stunned that they were able to find our house because we were always so cautious. However, we had built in a powerful self-destruct detonator into our laboratory just in case this ever happened. However, this was the last resort to keep everything for being discovered. It would take all three of our passwords in order to initiate the self-destruct. Once it was started there would be precisely 7 minutes until everything blows with the force of a 2 kiloton nuke.

We took turns entering our passwords, and once we finished the timer started 6:59, 6:58, 6:57 etc.

Erik hit the button to open the launch bay doors overhead. The doors are big & heavy and take over a minute to fully open. As the doors were opening, we ran over to the ship and up the ramp. Once we got inside, Erik quickly closed the door. Me and Abran ran over to the controls and immediately started the engines to get ready for takeoff.

Once the doors overhead were completely opened Abran gradually engaged the engines. We started to float up and through the launch bay doors. Within a matter of minutes, we were hovering about 300 feet above the ground. We were able to see everything through the video monitors. All the agents looked up to see us hovering above. The commanding agent pulled out a Megaphone and said to us, "You in the craft! We know who you are! And you can't escape from here! There's a squadron of fighters jets on their way here right now! They'll shoot you down if you try to flee this area! If you land right now and give yourselves up! You will not be arrested; in fact, we would like to make you some offers! The first offer is civil immunity, which means that no matter what crime you commit from this moment forward you will not be charged. In fact, it would be like it never happened. The second offer is high level, high paying jobs in the United States military where you will be handsomely rewarded!"

When I heard this, I thought [These guys must be crazy if they think we're gonna listen to them. I mean you can't trust **anyone** in the government, no matter how genuine they might sound!] Then I picked up our microphone to answer them saying, "Listen to me! This is extremely important! We've set our laboratory to self-destruct in 5 minutes! You need to get the hell outta here right now! And as for your offer goes, <u>we respectfully decline!</u>" The commanding agent pleaded with us saying, "Listen, we can sweeten the deal, just give us a chance!!!" I replied into the microphone saying, "Sorry no deal! However, I must remind you that you now have 4 minutes left! So, I suggest that you get moving NOW!!!"

Suddenly appearing on the horizon Abran saw small groups of military jets flying in from all sides. He immediately turned the shields on and began carefully monitoring them. The jets were about 30 miles away but closing fast! Once they were close enough they started firing side-winder missiles at us.

That's when Abran yelled, "They just fired at us! Let's get the HELL outta here!!!" I Immediately realized that Abran was right. Then I thought [We've wasted enough time bullshitting with these imbeciles. If they're too stupid to heed our warnings and end up dying! Then it's not my problem at least we gave them fair warning.] So, I yelled to Abran, **"PUNCH IT!!!"**

That's when Abran fully engaged the ship's engines and we blasted off so fast, that we left a crater over a 1000feet wide.

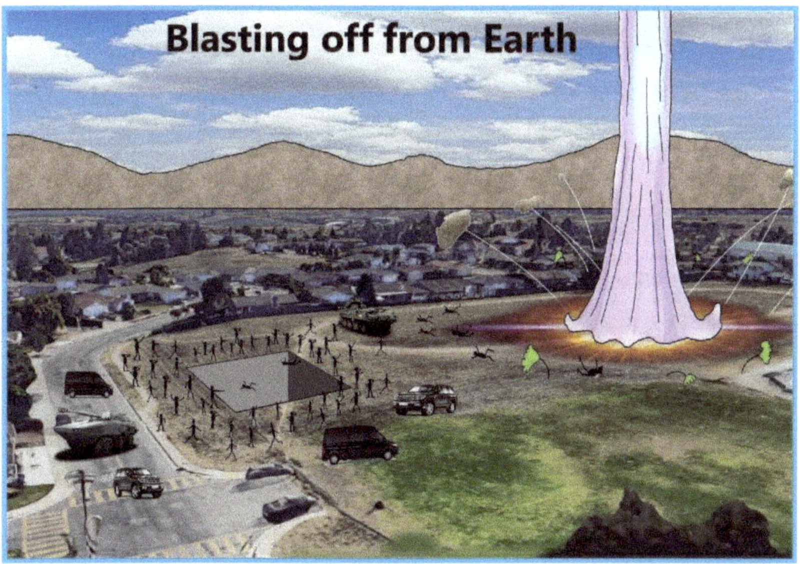

Within a few seconds we had reached the speed of Mach 200 [One hundred fifty-three thousand miles per hour] and still gaining speed. Our mother almost passed out with the shock of the whole experience. We all sat back and watched the Earth whizzing away from us at an ever-increasing speed. Right after we passed by the moon we detected that our laboratory had just exploded. After which I said, "I hope all those idiots got away from our laboratory before it blew up!"

The further we got from Earth I started to wonder. [After we explore, will we ever see the Earth again? What about our family and friends that are still behind on Earth, will we ever see them again? What if we find something out in space that's better than the Earth?]

During this whole time, I kept having the most intense déjà vu with the overwhelming feeling that we had rectified our situation.

We all felt a wave of contentment and happiness come over us. Somehow we knew that all was good now. After we had been traveling for a little while and everyone had relaxed. We revealed our secret plans to travel and explore the universe to our mother. After what she had witnessed she fully understood why everything had happened. She too was exhilarated and looking forward to exploring the universe with us.

Somehow we all knew that our amazing adventure was about to begin......

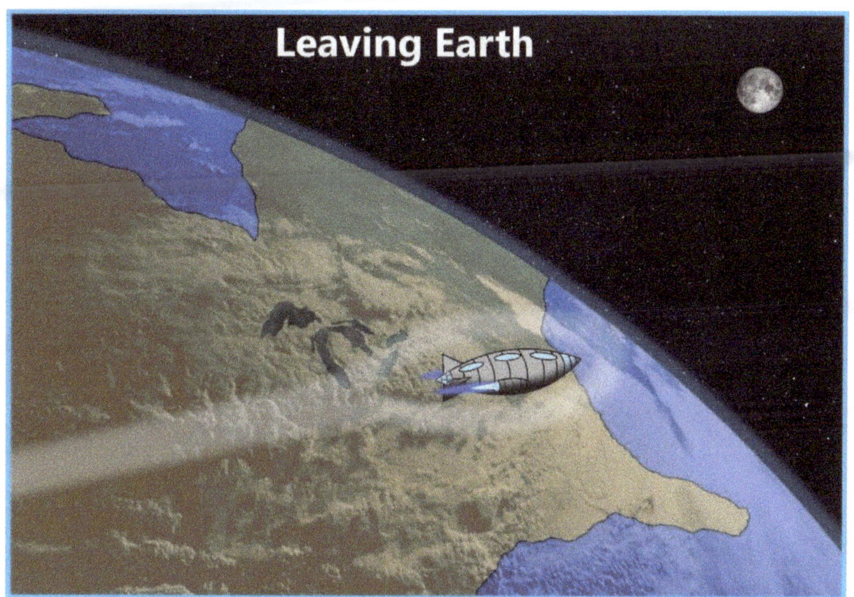

The End....

Maps

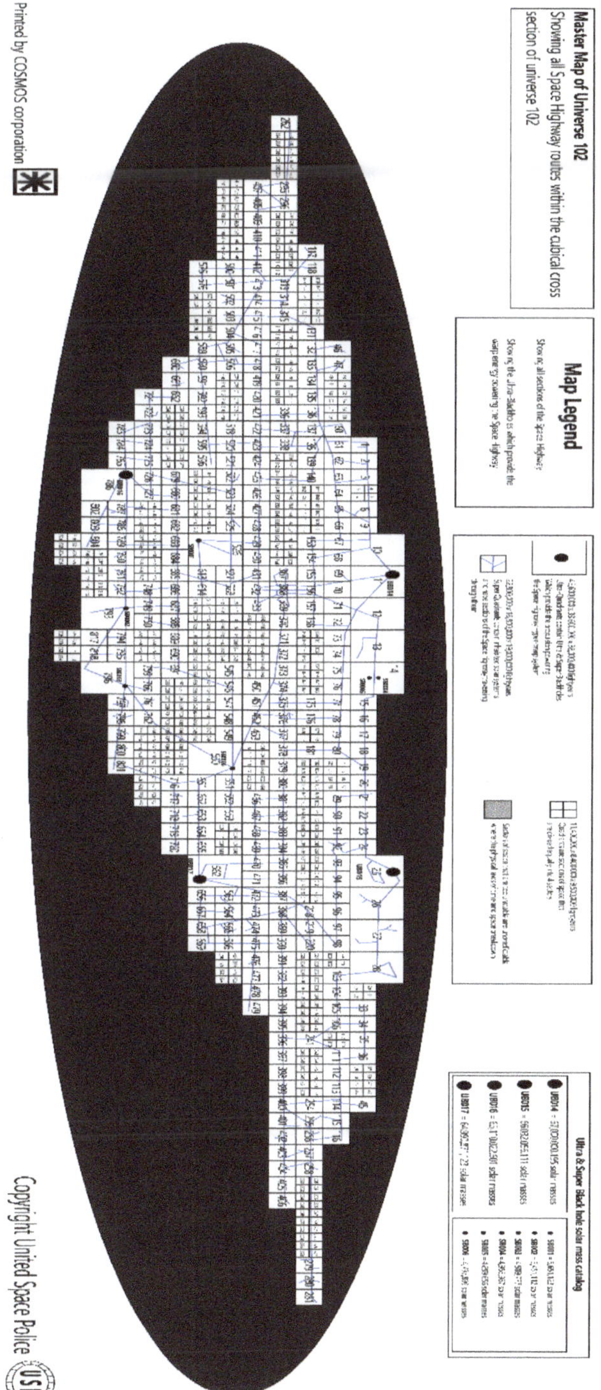
Master Map of Universe 102
Showing all Space Highway routes within the cubical cross section of universe 102
Printed by COSMOS corporation
Map Legend
Ultra & Super Black hole solar mass catalog
Copyright United Space Police

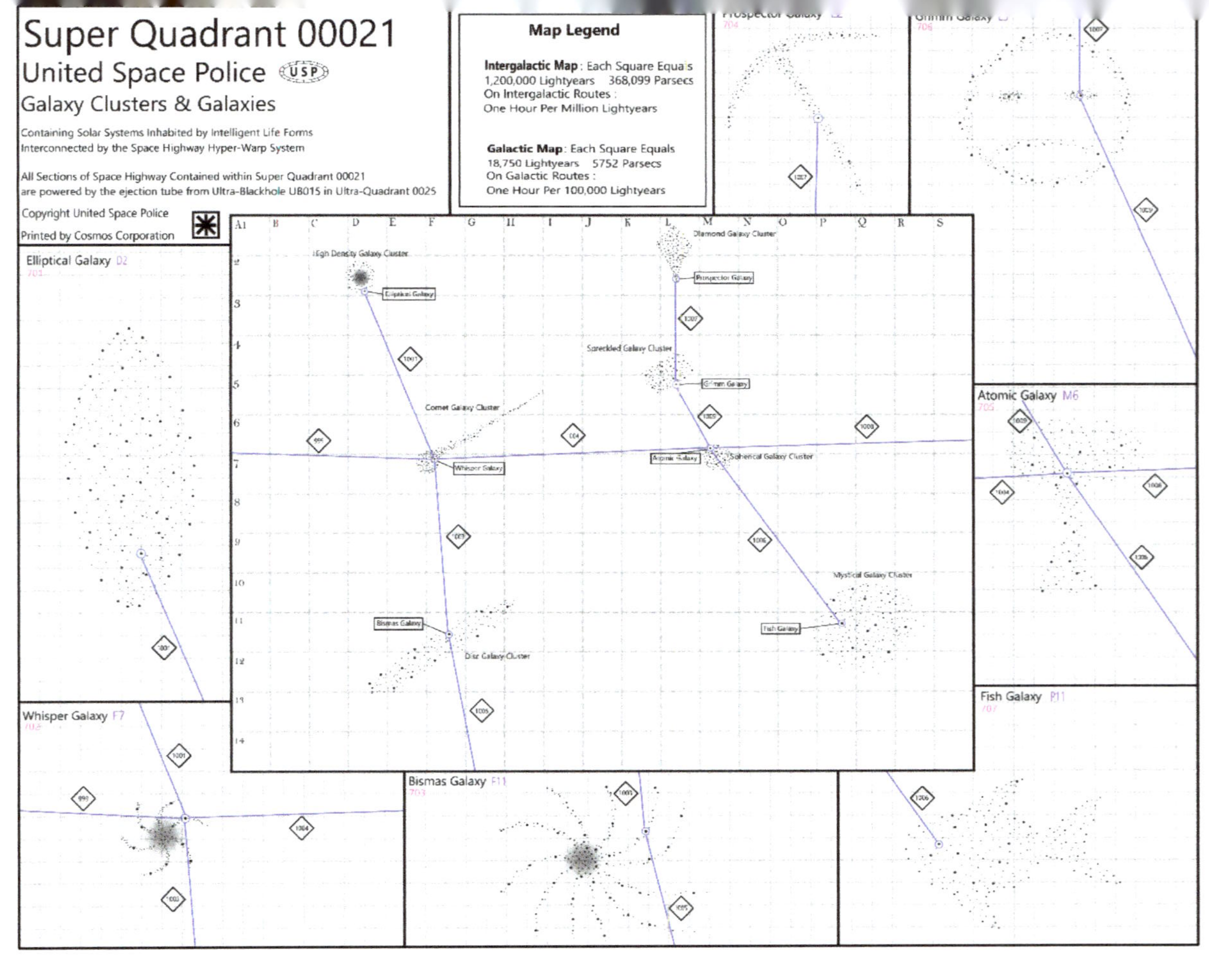

Super Quadrant 00021
United Space Police USP
Galaxy Clusters & Galaxies
Containing Solar Systems Inhabited by Intelligent Life Forms
Interconnected by the Space Highway Hyper-Warp System
All Sections of Space Highway Contained within Super Quadrant 00021
are powered by the ejection tube from Ultra-Blackhole UB015 in Ultra-Quadrant 0025
Copyright United Space Police
Printed by Cosmos Corporation
Map Legend
Intergalactic Map : Each Square Equals
1,200,000 Lightyears   368,099 Parsecs
On Intergalactic Routes :
One Hour Per Million Lightyears
Galactic Map : Each Square Equals
18,750 Lightyears   5752 Parsecs
On Galactic Routes :
One Hour Per 100,000 Lightyears
Elliptical Galaxy D2
Atomic Galaxy M6
Fish Galaxy P11
Whisper Galaxy F7
Bismas Galaxy F11
Prospector Galaxy
Grimm Galaxy
High Density Galaxy Cluster
Diamond Galaxy Cluster
Prospector Galaxy
Speckled Galaxy Cluster
Grimm Galaxy
Comet Galaxy Cluster
Whisper Galaxy
Atomic Galaxy
Spherical Galaxy Cluster
Elliptical Galaxy
Bismas Galaxy
Disc Galaxy Cluster
Mystical Galaxy Cluster
Fish Galaxy

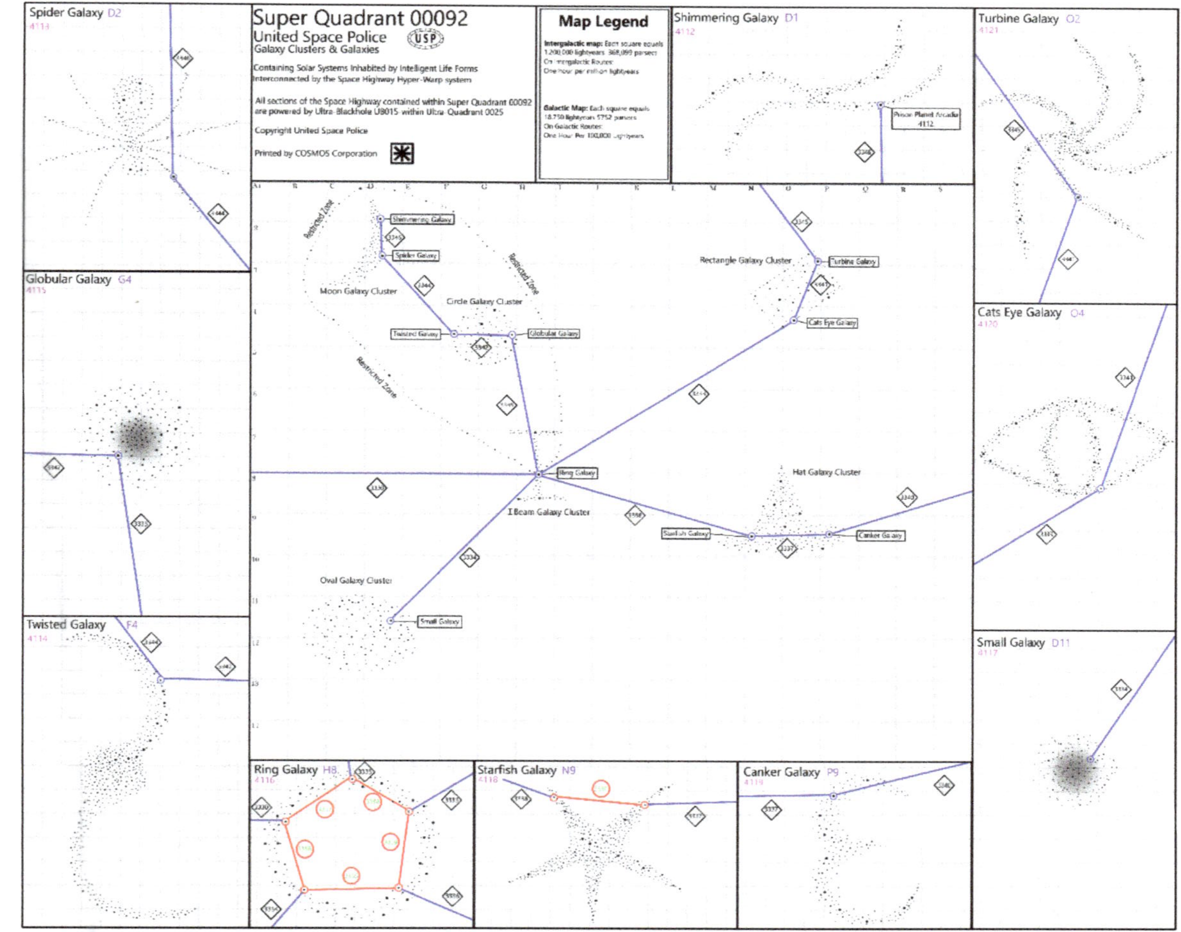

Spider Galaxy D2
4113
Globular Galaxy G4
4115
Twisted Galaxy F4
4114
Super Quadrant 00092
United Space Police
USP
Galaxy Clusters & Galaxies
Containing Solar Systems inhabited by Intelligent Life Forms
interconnected by the Space Highway Hyper-Warp system
All sections of the Space Highway contained within Super Quadrant 00092
are powered by Ultra-Blackhole U8015 within Ultra-Quadrant 0025
Copyright United Space Police
Printed by COSMOS Corporation
Map Legend
Intergalactic map: Each square equals
1,200,000 lightyears 368,099 parsecs
On Intergalactic Routes:
One Hour per million lightyears
Galactic Map: Each square equals
18,750 lightyears 5752 parsecs
On Galactic Routes:
One Hour Per 100,000 Lightyears
Shimmering Galaxy D1
4112
Turbine Galaxy O2
4121
Restricted Zone
Shimmering Galaxy
Spider Galaxy
Restricted Zone
Moon Galaxy Cluster
Circle Galaxy Cluster
Rectangle Galaxy Cluster
Turbine Galaxy
Twisted Galaxy
Globular Galaxy
Cats Eye Galaxy
Restricted Zone
Ring Galaxy
Hat Galaxy Cluster
I Beam Galaxy Cluster
Starfish Galaxy
Canker Galaxy
Oval Galaxy Cluster
Small Galaxy
Prison Planet Arcadia
4112
Globular Galaxy G4
4115
Cats Eye Galaxy Q4
4120
Twisted Galaxy F4
4114
Small Galaxy D11
4117
Ring Galaxy H8
4116
Starfish Galaxy N9
4118
Canker Galaxy P9
4119

Super Quadrant 00368
United Space Police   USP
Galaxy Clusters & Galaxies
Containing Solar Systems Inhabited by Intelligent life forms
Interconnected by the Space Highway Hyper-warp system.
Super Quadrant 00368 Space Highway system is powered by
the ejection tube of Ultra BlackHole UB014 within
Ultra-Quadrant 0011
Copyright United Space Police
Printed by COSMOS corporation
Neptunian Galaxy
Golden Galaxy
Spiral Galaxy
Blue Galaxy
The Great Galaxy
Star Galaxy
Fire Galaxy
Reptiless Galaxy
Micro Galaxy
Eagle Galaxy Cluster
Wolverine Galaxy Cluster
Local Group Galaxy Cluster
Spiral Galaxy Cluster
Serpent Galaxy Cluster
Celestrial Hex Galaxy Cluster
Crystal Galaxy Cluster
Globular Super Galaxy Cluster
Triangle Galaxy Cluster
Map Legend
Intergalactic Map: Each square equals
1,200,000 Lightyears   368,099 Parsecs
On Intergalactic Routes :
One Hour Per Million Lightyears
Galactic Map: Each square equals
18,750 lightyears   5752 Parsecs
On Galactic Routes :
One Hour Per 100,000 Lightyears

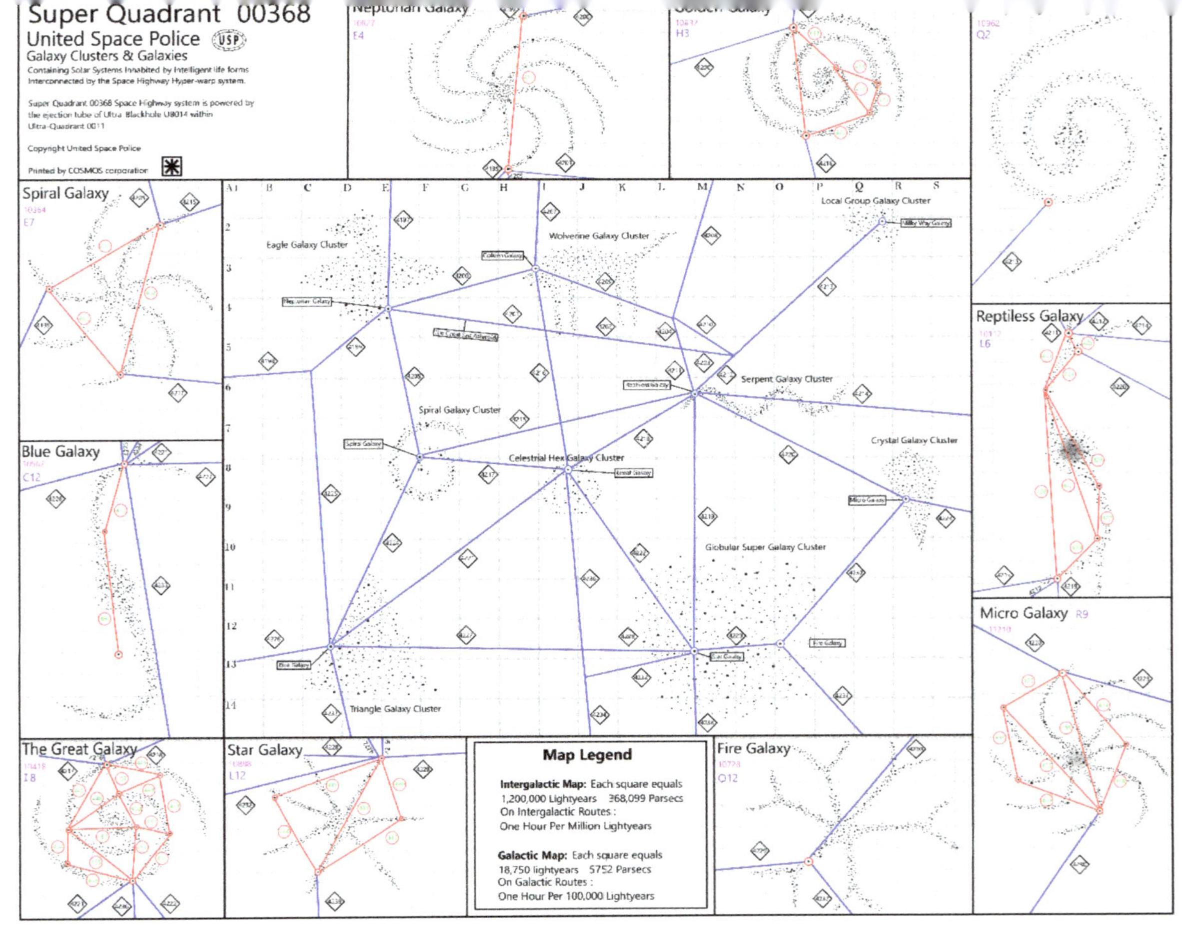

254

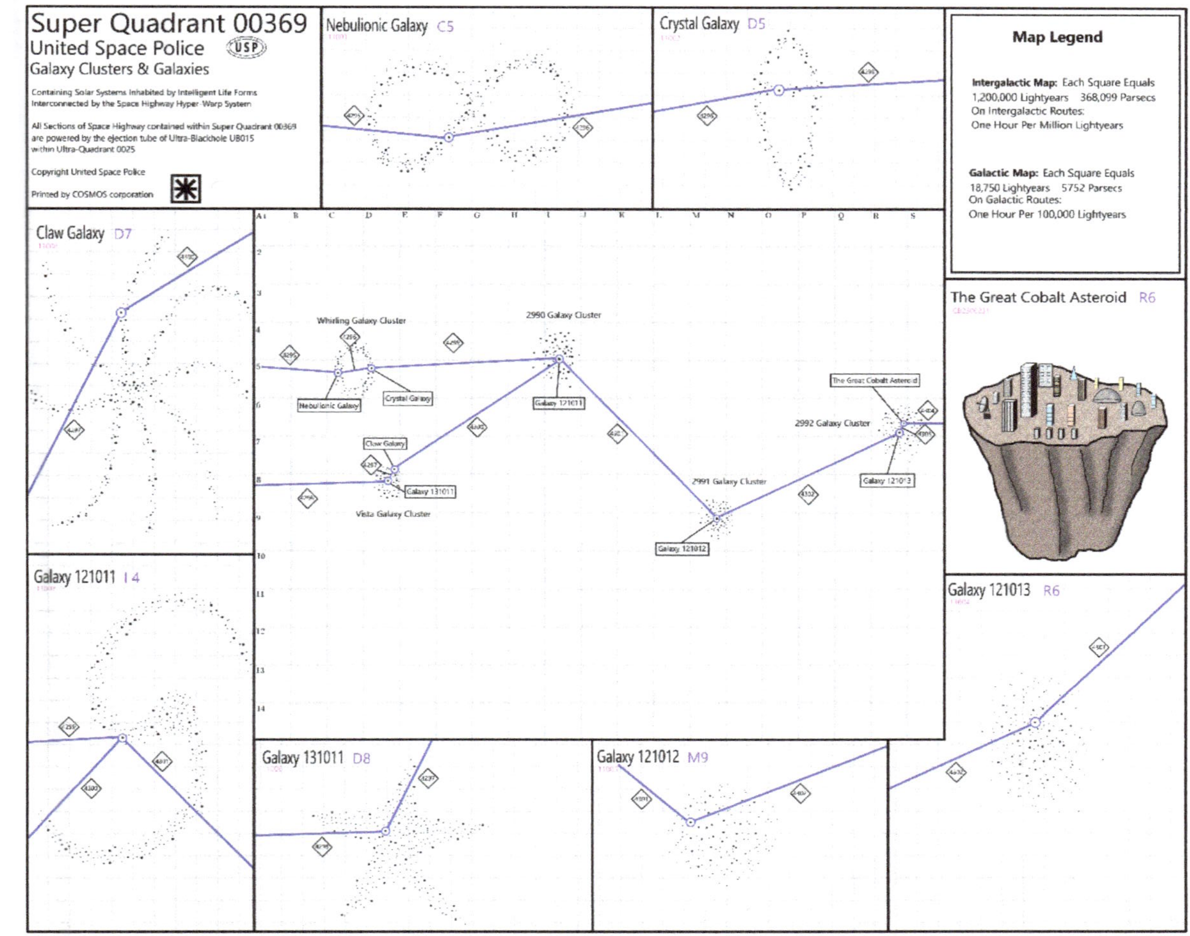

Super Quadrant 00369
United Space Police   USP
Galaxy Clusters & Galaxies
Containing Solar Systems Inhabited by Intelligent Life Forms
Interconnected by the Space Highway Hyper-Warp System
All Sections of Space Highway contained within Super Quadrant 00369 are powered by the ejection tube of Ultra-Blackhole UB015 within Ultra-Quadrant 0025
Copyright United Space Police
Printed by COSMOS corporation
Nebulionic Galaxy   C5
Crystal Galaxy   D5
Map Legend
Intergalactic Map: Each Square Equals 1,200,000 Lightyears   368,099 Parsecs
On Intergalactic Routes:
One Hour Per Million Lightyears
Galactic Map: Each Square Equals 18,750 Lightyears   5752 Parsecs
On Galactic Routes:
One Hour Per 100,000 Lightyears
Claw Galaxy   D7
Whirling Galaxy Cluster
2990 Galaxy Cluster
Nebulionic Galaxy
Crystal Galaxy
Galaxy 121011
Claw Galaxy
Galaxy 131011
Vista Galaxy Cluster
The Great Cobalt Asteroid
2992 Galaxy Cluster
2991 Galaxy Cluster
Galaxy 121013
Galaxy 121012
The Great Cobalt Asteroid   R6
Galaxy 121011   I4
Galaxy 121013   R6
Galaxy 131011   D8
Galaxy 121012   M9

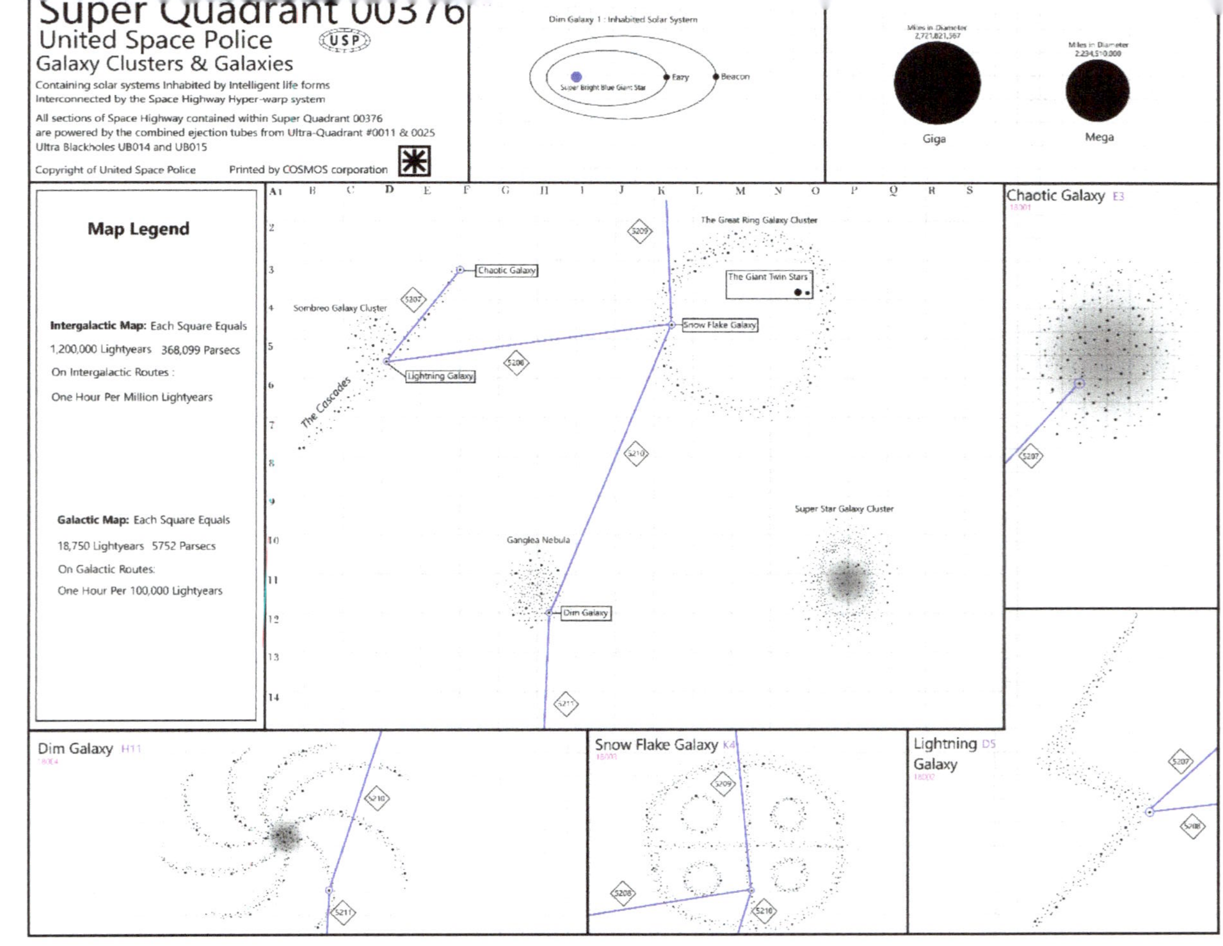
Super Quadrant 00376
United Space Police
USP
Galaxy Clusters & Galaxies
Containing solar systems Inhabited by Intelligent life forms
Interconnected by the Space Highway Hyper-warp system
All sections of Space Highway contained within Super Quadrant 00376
are powered by the combined ejection tubes from Ultra-Quadrant #0011 & 0025
Ultra Blackholes UB014 and UB015
Copyright of United Space Police   Printed by COSMOS corporation

Dim Galaxy 1 : Inhabited Solar System
Super bright Blue Giant Star
Eazy
Beacon

Miles in Diameter
2,721,821,567
Giga

Miles in Diameter
2,234,510,000
Mega

Map Legend

Intergalactic Map: Each Square Equals
1,200,000 Lightyears   368,099 Parsecs
On Intergalactic Routes :
One Hour Per Million Lightyears

Galactic Map: Each Square Equals
18,750 Lightyears 5752 Parsecs
On Galactic Routes:
One Hour Per 100,000 Lightyears

The Great Ring Galaxy Cluster
The Giant Twin Stars
Sombreo Galaxy Cluster
The Cascodes
Chaotic Galaxy
Snow Flake Galaxy
Lightning Galaxy
Ganglea Nebula
Super Star Galaxy Cluster
Dim Galaxy

Chaotic Galaxy E3

Dim Galaxy H11

Snow Flake Galaxy K4

Lightning D5
Galaxy

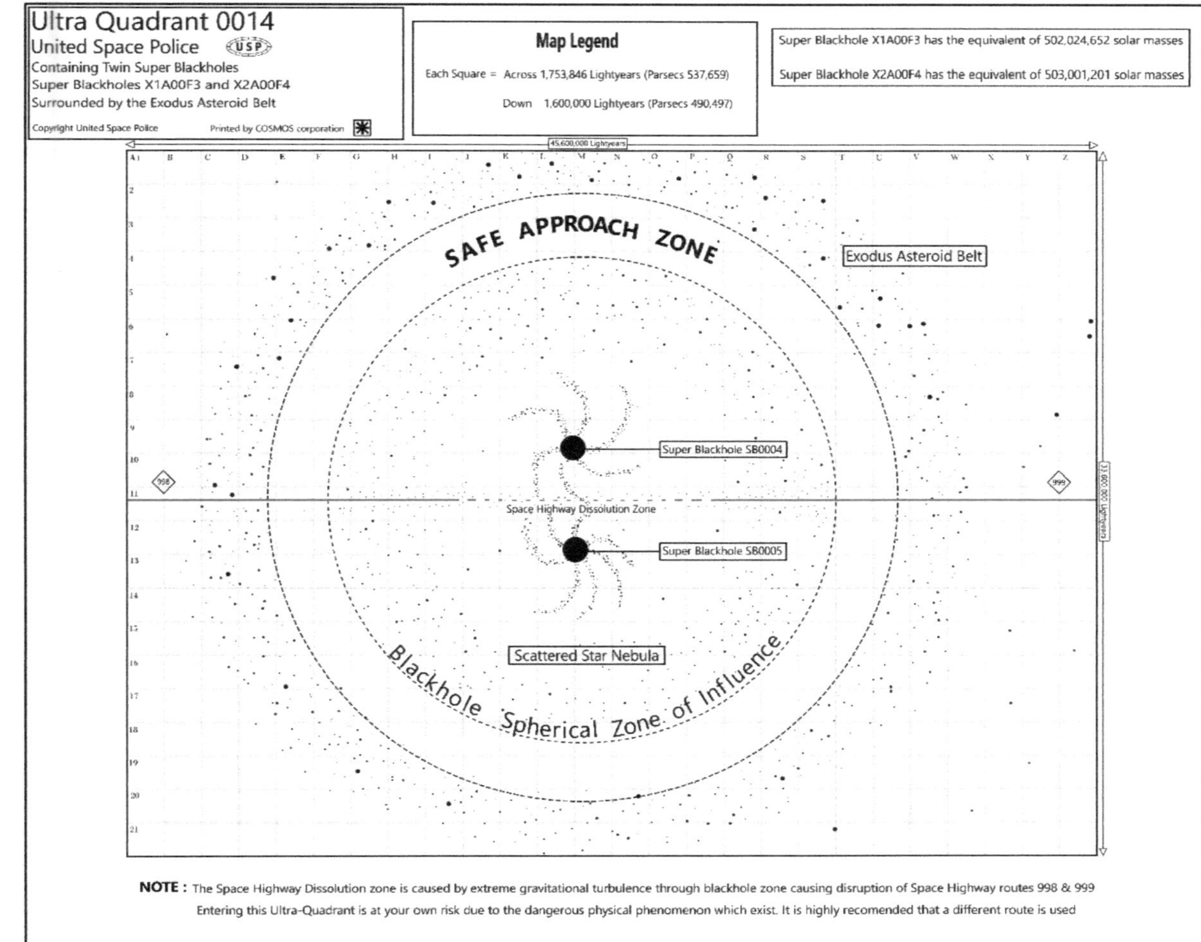

NOTE : The Space Highway Dissolution zone is caused by extreme gravitational turbulence through blackhole zone causing disruption of Space Highway routes 998 & 999

Entering this Ultra-Quadrant is at your own risk due to the dangerous physical phenomenon which exist. It is highly recomended that a different route is used

Narrator: Epilogue

Even though this story has ended it hasn't really ended and to say that about any story would be un-true. Like all stories either fact or fiction mimic a true part of the human experience. As we have all experienced: Joy , Pain, Fear, Love, and Hate.

I started drafting this book over 20 years before it's completion. I played this childhood game with my cousin, brothers and by myself. It has been inspired by several movies and television shows.

The reason why I decided to finish this book is not really known to me. Other than the fact that I had just went through the worst experience in my life. After which I barely survived! The utter emotional distress and despair to the point of suicide! The divine force which compelled me to finish this book will hopefully inspire all who read it. To finish whatever project or lifelong dream that you've always wanted to complete. You have to believe in yourself and persevere at all costs and be willing to put your life on the line!

If I inspire one person then I know I have succeeded.

May all your childhood dreams live on.

Farewell my friend......

www.ingramcontent.com/pod-product-compliance
Lightning Source LLC
Chambersburg PA
CBHW041046310726
48978CB00011BA/444